Erwin Dee Kord (Ed.)

Waltenhausen

Erwin Dee Kord (Ed.)

Waltenhausen

Municipalities of Germany, Regierungsbezirk, Miesbach, Bad Wiessee

Solv

Imprint

Permission is granted to copy, distribute and/or modify this document under the terms of the GNU Free Documentation License, Version 1.2 or any later version published by the Free Software Foundation; with no Invariant Sections, with the Front-Cover Texts, and with the Back- Cover Texts. A copy of the license is included in the section entitled "GNU Free Documentation License".

All parts of this book are extracted from Wikipedia, the free encyclopedia (www.wikipedia.org).

You can get detailed informations about the authors of this collection of articles at the end of this book. The editors (Ed.) of this book are no authors. They have not modified or extended the original texts.

Pictures published in this book can be under different licences than the GNU Free Documentation License. You can get detailed informations about the authors and licences of pictures at the end of this book.

The content of this book was generated collaboratively by volunteers. Please be advised that nothing found here has necessarily been reviewed by people with the expertise required to provide you with complete, accurate or reliable information. Some information in this book maybe misleading or wrong. The Publisher does not guarantee the validity of the information found here. If you need specific advice (f.e. in fields of medical, legal, financial, or risk management questions) please contact a professional who is licensed or knowledgeable in that area.

Any brand names and product names mentioned in this book are subject to trademark, brand or patent protection and are trademarks or registered trademarks of their respective holders. The use of brand names, product names, common names, trade names, product descriptions etc. even without a particular marking in this works is in no way to be construed to mean that such names may be regarded as unrestricted in respect of trademark and brand protection legislation and could thus be used by anyone.

Cover image: www.ingimage.com
Concerning the licence of the cover image please contact ingimage.

Publisher:
Solv is a trademark of
International Book Market Service Ltd., 17 Rue Meldrum, Beau Bassin, 1713-01 Mauritius
Email: info@bookmarketservice.com
Website: www.bookmarketservice.com

Published in 2012

Printed in: U.S.A., U.K., Germany. This book was not produced in Mauritius.

ISBN: 978-613-8-64567-2

Contents

Articles

References

Waltenhausen

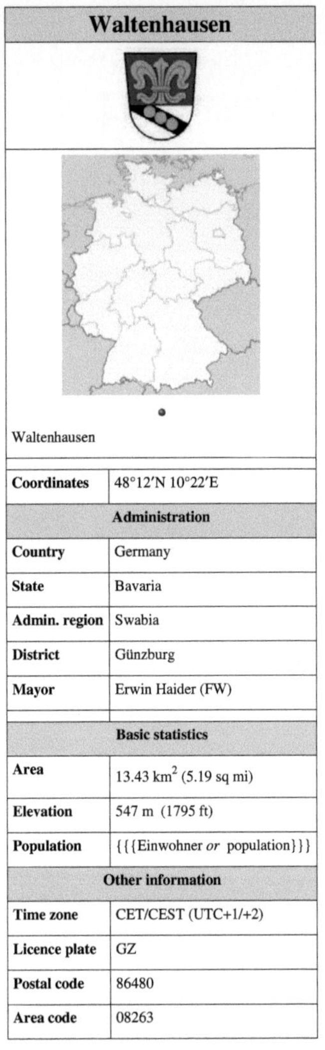

Waltenhausen	
Waltenhausen	
Coordinates	48°12′N 10°22′E
Administration	
Country	Germany
State	Bavaria
Admin. region	Swabia
District	Günzburg
Mayor	Erwin Haider (FW)
Basic statistics	
Area	13.43 km^2 (5.19 sq mi)
Elevation	547 m (1795 ft)
Population	{{{Einwohner *or* population}}}
Other information	
Time zone	CET/CEST (UTC+1/+2)
Licence plate	GZ
Postal code	86480
Area code	08263

Waltenhausen is a municipality in the district of Günzburg in Bavaria in Germany.

References

Municipalities of Germany

Municipalities (*Gemeinde*) are the lowest level of territorial division **in Germany**. This may be the fourth level of territorial division in Germany, apart from those states which include *Regierungsbezirke* (singular: *Regierungsbezirk*), where municipalities then become the fifth level.

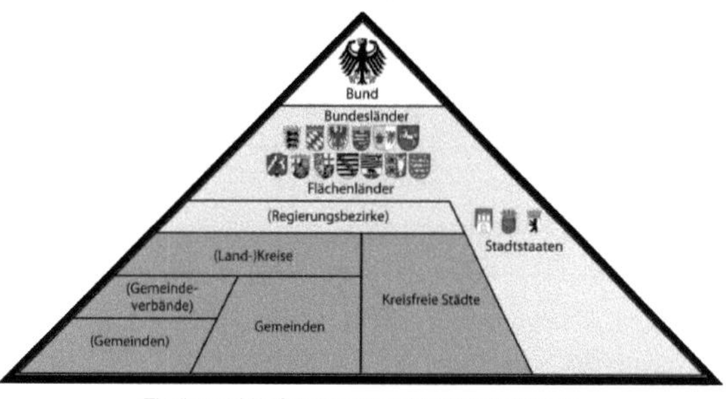

The "pyramid" of German administrative subdivisions

Overview

With more than 3,400,000 inhabitants, the most populated one is the city of Berlin; and the least populated is Wiedenborstel (8 inhabitants in 2010), in Schleswig-Holstein

Municipalities per federal state

List updated at August 1, 2009.

Federal state	Municipalities	M. with city status	Average nr. of inhabitants[1]	Average surface (km²)[2]	List (Cities, Towns, Municipalities)
Baden-Württemberg	1,101	312	9,764	32.41	C, T, M
Bavaria	2,056	315	6,090	33.03	C, T, M
Berlin	1	1	3,416,255	891.02	Berlin
Brandenburg	419	112	6,052	70.36	C, T, M
Bremen	2	2	331,541	202.14	Bremen, Bremerhaven
Hamburg	1	1	1,770,629	755.16	Hamburg
Hesse	426	189	14,255	48.80	C, T, M
Lower Saxony	1,022	164	7,800	45.25	C, T, M
Mecklenburg-Vorpommern	818	84	2,053	28.34	C, T, M
North Rhine-Westphalia	396	268	45,446	86.08	C, T, M
Rhineland-Palatinate	2,306	123	1,754	8.61	C, T, M
Saarland	52	17	19,935	49.40	C, T, M
Saxony	491	178	8,595	37.51	C, T, M
Saxony-Anhalt	851	118	2,835	24.03	C, T, M
Schleswig-Holstein	1,116	63	2,542	14.07	C, T, M
Thuringia	955	126	2,397	16.93	C, T, M
Germany	**12,013**	**2,073**	**6,844**	**29.35**	**C, T, M**

References

[1]   Source: Statistisches Bundesamt (as of December 31, 2007)

[2]   Source: Statistisches Bundesamt (as of December 31, 2006), for the computation less the municipality-free areas

See also

- List of municipalities in Germany

Regierungsbezirk

In Germany, a **Government District**, in German: *Regierungsbezirk* – pronounced [ʁeˈgiːʁʊŋsbəˌtsɪʁk] is a subdivision of certain federal states (*Bundesländer*).

They are above the *Kreise*, *Landkreise*, and *kreisfreie Städte*.[1] The *Regierungsbezirk* is governed by a *Bezirksregierung* (literally "district government") and led by a *Regierungspräsident* (literally "government president").

History

The first *Regierungsbezirke* were created by the Kingdom of Prussia, which divided its provinces into 25 *Regierungsbezirke* in 1808/1816. The *Regierungsbezirke* of North Rhine-Westphalia are in direct continuation of those created in 1815. Other states of the German Empire created similar entities, named *Kreishauptmannschaft* (in Saxony) or *Kreis* (in Bavaria and Württemberg) (not to be confused with the *Kreis* or *Landkreis* today). During the Third Reich, the Nazi government unified the naming; since then all these entities are called *Regierungsbezirk*.

Regierungsbezirke as from 1st of August 2008. In the map there are also shown the former RBs of Lower Saxony, Rhineland-Palatinate and Saxony-Anhalt

On January 1, 2000 Rhineland-Palatinate disbanded its three *Regierungsbezirke* Koblenz, Rheinhessen-Pfalz and Trier - the employees and assets of the three *Bezirksregierungen* were converted into three public authorities responsible for the whole state, each covering a part of the former responsibilities of the *Bezirksregierung*.

On January 1, 2004, Saxony-Anhalt disbanded its three *Regierungsbezirke*: Dessau, Halle and Magdeburg. The responsibilities are now covered by a *Landesverwaltungsamt* with three offices at the former seats of the *Bezirksregierungen*.

At the foundation of Lower Saxony in 1946 by the merger of the three former Free States of Brunswick, Oldenburg, Schaumburg-Lippe and the former Prussian province of Hanover the former two states became *Verwaltungsbezirke* (roughly administrative regions of extended competence) within Lower Saxony besides the less autonomous Prussian-style *Regierungsbezirke* comprising the former Province of Hanover and the tiny Schaumburg-Lippe. These differences were levelled on 1 January 1978, when four territorially redeployed Regierungsbezirke replaced the two Verwaltungsbezirke and the old six Regierungsbezirke: Brunswick and Oldenburg as well as Aurich, Hanover (remaining mostly the same), Hildesheim, Lüneburg (old), Osnabrück and Stade. On January 1, 2005, Lower Saxony disbanded its remaining four Regierungsbezirke: Braunschweig, Hanover, Lüneburg, and Weser-Ems.

In 2005, North Rhine-Westphalia planned to abolish its five *Regierungsbezirke* and create three self-government entities. The old, "Prussian-style", Regierungsbezirk had no self-government organs.

On August 1, 2008, Saxony restructured its districts (*Landkreise*) and changed the name of its *Regierungsbezirke* to *Direktionsbezirke*. This was necessary because one of the new districts did not fit with the borders of the old *Regierungsbezirke* and some responsibilities are now covered by the districts. The *Direktionsbezirke* are still named *Chemnitz, Dresden* and *Leipzig*. The authorities of the *Direktionsbezirke* are named *Landesdirektion* and their Presidents are called *Präsident der Landesdirektion* instead of *Regierungspräsidium* and *Regierungspräsident*.

Local existence

Not all *Bundesländer* have this subdivision; some are directly divided into districts. Currently, five states are divided into 22 *Regierungsbezirke*, ranging in population from 5,255,000 (Düsseldorf) to 1,065,000 (Gießen):

- Baden-Württemberg: Freiburg, Karlsruhe, Stuttgart, Tübingen
- Bavaria: Upper Bavaria, Lower Bavaria, Upper Palatinate, Upper Franconia, Middle Franconia, Lower Franconia, Swabia
- Hesse: Darmstadt, Gießen, Kassel
- North Rhine-Westphalia: Arnsberg, Cologne, Detmold, Düsseldorf, Münster
- Saxony: Chemnitz, Dresden, Leipzig

The regional governments of the *Regierungsbezirke* are mostly concerned with administrative decisions on a regional level. Most administrative routines are handled by the municipal government (of the city or the *Kreis*) while legislation is passed by the parliament of the *Bundesland* or on a national level.

References

[1] Regional Governments in France, Germany, Poland and The Netherlands (HTML version of PowerPoint presentation) (http://google.com/ search?q=cache:I1oTRcSOTJ8J:www.hanse-passage.net/hansepassage/hpcms/uploads/_presentation%20analysis.ppt+"regierungsbezirk+ is"&hl=en&ct=clnk&cd=18&lr=lang_en) - Cachet, A (coordinator), Erasmus University, Rotterdam

Bad Wiessee

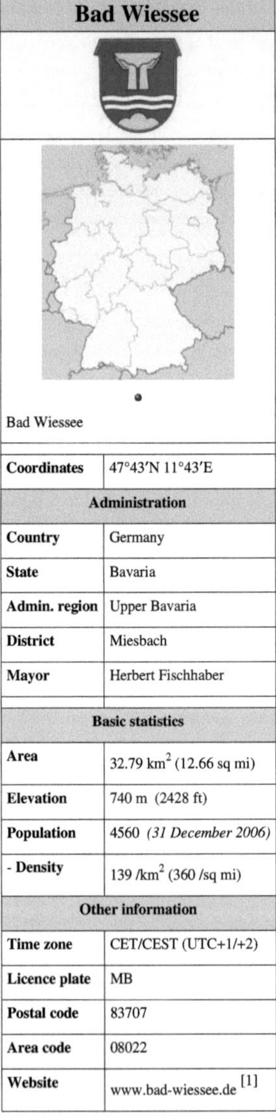

Bad Wiessee	
Coordinates	47°43′N 11°43′E
Administration	
Country	Germany
State	Bavaria
Admin. region	Upper Bavaria
District	Miesbach
Mayor	Herbert Fischhaber
Basic statistics	
Area	32.79 km^2 (12.66 sq mi)
Elevation	740 m (2428 ft)
Population	4560 *(31 December 2006)*
- Density	139 /km^2 (360 /sq mi)
Other information	
Time zone	CET/CEST (UTC+1/+2)
Licence plate	MB
Postal code	83707
Area code	08022
Website	www.bad-wiessee.de [1]

Bad Wiessee

Bad Wiessee is a spa town on Lake Tegernsee, Bavaria, Germany. The name "Bad" means for "spa" or "baths", while "Wiessee" derives from "West See", meaning "western part of the lake".

Bad Wiessee is known for its healing sulfur-fountain, discovered by the Dutch oil explorer Adriaan Stoop.[2] People spend their holidays in Bad Wiessee because of its quiet atmosphere and its location at the north side of the Alps.

Bad Wiessee is also notorious as the scene of the key events within the Night of the Long Knives, June 30, 1934, when Hitler and the Schutzstaffel (SS) purged the leadership of the Sturmabteilung (SA), many of whom were staying at the resort.

References

[1] http://www.bad-wiessee.de/

[2] History of Bad Wiessee (http://www.live-like-a-german.com/destinations/show/ bad-wiessee/)

Maria Himmelfahrt Church in Bad Wiessee.

External links

- Official website (http://www.bad-wiessee.de)
- Lake tegernsee and its towns (http://www.treffpunkt-tegernsee.de).

Miesbach (district)

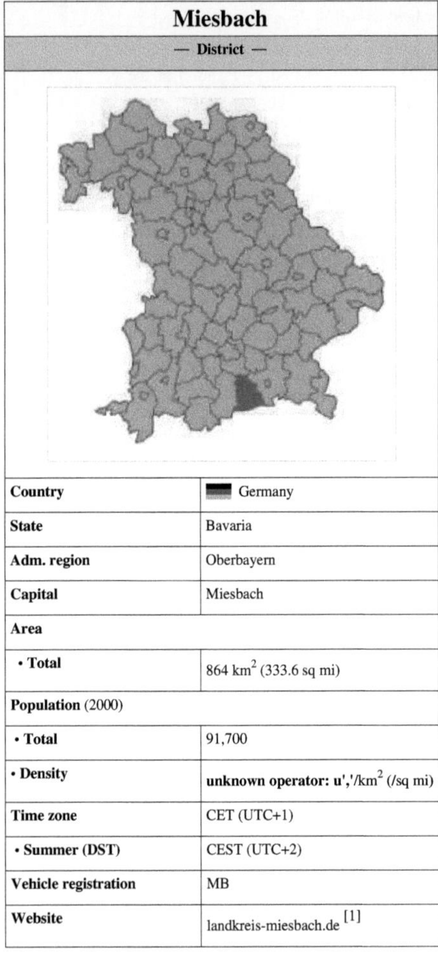

Miesbach	
— District —	
Country	Germany
State	Bavaria
Adm. region	Oberbayern
Capital	Miesbach
Area	
• Total	864 km^2 (333.6 sq mi)
Population (2000)	
• Total	91,700
• Density	unknown operator: u','/km^2 (/sq mi)
Time zone	CET (UTC+1)
• Summer (DST)	CEST (UTC+2)
Vehicle registration	MB
Website	landkreis-miesbach.de [1]

Miesbach is a district in Bavaria, Germany. It is bounded by (from the west and clockwise) the districts of Bad Tölz-Wolfratshausen, Munich and Rosenheim, and by the Austrian state of Tyrol.

History

In medieval times most of the district was occupied by clerical states. The Miesbach district is the union of the areas that were formerly occuped by the Hohenwaldeck county, the territories owned by the powerful Tegernsee Abbey, the territories owned by the Weyarn Abbey and Valley County. Hohenwaldeck was annexed by Bavaria in 1734, Valley in 1777. The clerical states were dissolved in 1803 and fell to Bavaria as well.

Miesbach was established in 1803 the foundation ceremony took place in the court district of Hohenwaldeck. In 1818 Tegernsee was established. The same year the Bad Aibling district was established too and Miesbach had to deliver 12 municipalities. In 1939 Tegernsee was merged into Miesbach. During the territorial reform in Bavaria in 1972 Otterffing was merged into Miesbach too and the district Wolfratshausen as well.

Geography

The southern half of the district is located in the Bavarian Alps. This section of the Alps is called Mangfall Mountains (*Mangfallgebirge*) as the River Mangfall has its source here. The most important peaks are **Rotwand** ("Red Wall", 1884 m), the highest mountain in the area and Wendelstein (1936 m) .

The Mangfall River, a short but broad affluent of the Inn River, flows northward with the district's municipalities on either bank. In the initial part of its course it runs through **Tegernsee** Lake (9 km²).

Coat of arms

The coat of arms displays:
- a red eagle and two crossed staffs, symbolising the county of Hohenwaldeck
- two water lily leaves above a wavy line, the ancient arms of the monastery of Tegernsee

Government

Charge	Name	Years in office
District Minister	Franz Xaver Steyrer	1803–1806
	Max Graf von Preysing	1806–1813
	Joseph Wiesend	1813–1846
	Wolfgang Knorr	1846–1848
	Alois Schmid	1848–1851
	Carl Bollweg	1851–1862, 1874
Joint President	Carl Bollweg	1851, 1862–1874
	Ludwig Mayr	1874–1888
	Uhl	1888–1890
	Carl Riezler	1890–1911
	Bernhard de Rudder	1911–1929
	Hermann Kopp	1929–1938, 1943
District Administrator	Kemnitzer	1943–1944
	Dr. Frick	1944–1945
	Dr. von Wehner	May 10, 1945 – August 1, 1945
	Schindler	August 1, 1945 – May 10, 1946
	Friedrich Roith	May 11, 1946 – May 27, 1946
	Freiherr von Schoen	May 28, 1946 – June 11, 1946
	Friedrich Roith	June 11, 1946 – September 25, 1946
	Dr. Süß	September 26, 1946 – May 31, 1948
	Meissner	June 5, 1948 – December 14, 1948
	Dr. Simon Beck	1949–1951
	Anton Bauer	1952–1955
	Dr. Walter	1955–1972
	Königsdorfer	1972–1987
	Wolfgang Gröbl	1987–2008
	Norbert Kerkel	since 2008
	Jakob Kreidl	

Towns and municipalities

Towns	Municipalities	
• Miesbach	• Bad Wiessee	• Kreuth
• Tegernsee	• Bayrischzell	• Otterfing
	• Fischbachau	• Rottach-Egern
	• Gmund am Tegernsee	• Schliersee
	• Hausham	• Valley
	• Holzkirchen	• Waakirchen
	• Irschenberg	• Warngau
		• Weyarn

References

[1] http://www.landkreis-miesbach.de

External links

• Official website (http://www.landkreis-miesbach.de/) (German)

Bavaria

Free State of Bavaria Freistaat Bayern	
— State of Germany —	
Flag	Coat of arms
Coordinates: 48°46′39″N 11°25′52″E	
Country	Germany
Capital	Munich
Government	
• Minister-President	Horst Seehofer (CSU)
• Governing parties	CSU / FDP
• Votes in Bundesrat	6 (of 69)
Area	
• Total	70549.44 km^2 (27239.3 sq mi)
Population (2010-11)[1]	
• Total	12542365
• Density	177.8/km^2 (460.5/sq mi)
Time zone	CET (UTC+1)
• Summer (DST)	CEST (UTC+2)
ISO 3166 code	DE-BY
GDP/ Nominal	€ 442.39 billion (2010) [2]
GDP per capita	€ 33,200 (2006)

NUTS Region	DE2
Website	bayern.de [3]

Bavaria, formally the **Free State of Bavaria** (German: *Freistaat Bayern*, pronounced [ˈfʁaɪʃtaːt ˈbaɪ.ɐn] (◀ listen)) is a state of Germany, located in the southeast of Germany. With an area of 70548 square kilometres (27200 sq mi), it is the largest state by area, forming almost 20% of the total land area of Germany. Bavaria is Germany's second most populous state (after North Rhine-Westphalia) with almost 12.5 million inhabitants, more than any of the three sovereign states on its borders. Bavaria's capital is Munich.

One of the oldest states of Europe, it was established as a duchy in the mid first millennium. In the 17th century, the Duke of Bavaria became a Prince-elector of the Holy Roman Empire. The Kingdom of Bavaria existed from 1806 to 1918, and Bavaria has since been a free state (republic). Modern Bavaria also includes parts of the historical regions of Franconia and Swabia.

History

The Bavarians emerged in a region north of the Alps, originally inhabited by the Gauls, which had been part of the Roman provinces of Raetia and Noricum. The Bavarians spoke Old High German but, unlike other Germanic groups, probably did not migrate from elsewhere (however, according to the narrative traditions collected by Anno, the Bishop of Cologne, and some other documents, the Bavarians had come from Armenia, the 'land of Noah's arc'[4]). Rather, they seem to have coalesced out of other groups left behind by Roman withdrawal late in the 5th century. These peoples may have included the Celtic Boii, some remaining Romans, Marcomanni, Allemanni, Quadi, Thuringians, Goths, Scirians, Rugians, Heruli. The name "Bavarian" ("Baiuvarii") means "Men of Baia" which may indicate

Prehistoric *Heunischenburg*, in the vicinity of Kronach

Bohemia, the homeland of the Celtic Boii and later of the Marcomanni. They first appear in written sources c. 520. Saint Boniface completed the people's conversion to Christianity in the early-8th century. Bavaria was, for the most part, unaffected by the Protestant Reformation.

From about 554 to 788, the house of Agilolfing ruled the Duchy of Bavaria, ending with Tassilo III who was deposed by Charlemagne.

Three early dukes are named in Frankish sources: Garibald I may have been appointed to the office by the Merovingian kings and married the Lombard princess Walderada when the church forbade her to King Chlothar I in 555. Their daughter, Theodelinde, became Queen of the Lombards in northern Italy and Garibald was forced to flee to her when he fell out with his Frankish overlords. Garibald's successor, Tassilo I, tried unsuccessfully to hold the eastern frontier against the expansion of Slavs and Avars around 600. Tassilo's son Garibald II seems to have achieved a balance of power between 610 and 616.

After Garibald II little is known of the Bavarians until Duke Theodo I, whose reign may have begun as early as 680. From 696 onwards he invited churchmen from the west to organize churches and strengthen Christianity in his duchy (it is unclear what Bavarian religious life consisted of before this time). His son, Theudebert, led a decisive Bavarian campaign to intervene in a succession dispute in the Lombard Kingdom in 714, and married his sister Guntrud to the Lombard King Liutprand. At Theodo's death the duchy was divided among his sons, but reunited under his grandson Hucbert.

At Hucbert's death (735) the duchy passed to a distant relative named Odilo, from neighbouring Alemannia (modern southwest Germany and northern Switzerland). Odilo issued a law code for Bavaria, completed the process of church organisation in partnership with St. Boniface (739), and tried to intervene in Frankish succession disputes by fighting for the claims of the Carolingian Grifo. He was defeated near Augsburg in 743 but continued to rule until his death in 748.

Middle Ages

Further information: Duchy of Bavaria

Tassilo III (b. 741 - d. after 794) succeeded his father at the age of eight after an unsuccessful attempt by Grifo to rule Bavaria. He initially ruled under Frankish oversight but began to function independently from 763 onwards. He was particularly noted for founding new monasteries and for expanding eastwards, fighting Slavs in the eastern Alps and along the River Danube and colonising these lands. After 781, however, his cousin Charlemagne began to pressure Tassilo to submit and finally deposed him in 788. The deposition was not entirely legitimate. Dissenters attempted a coup against

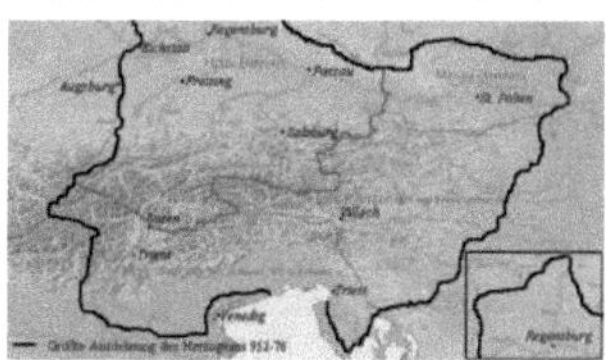

Bavaria in the 10th century

Charlemagne at Tassilo's old capital of Regensburg in 792, led by his own son Pépin the Hunchback (Pronounced *Pippin*), and the king had to drag Tassilo out of imprisonment to formally renounce his rights and titles at the Assembly of Frankfurt in 794. This is the last appearance of Tassilo in the sources and he probably died a monk. As all of his family were also forced into monasteries, this was the end of the Agilolfing dynasty.

For the next 400 years numerous families held the duchy, rarely for more than three generations. With the revolt of duke Henry the Quarrelsome in 976, Bavaria lost large territories in the south and south east. Among them a mark called "Ostarrichi" which was elevated to a duchy out of own right and given to the Babenberger family. This event marks the birth of Austria. The last, and one of the most important, of these dukes was Henry the Lion of the house of Welf, founder of Munich, de facto the second most powerful man in the empire as the ruler of two duchies. When in 1180, Henry the Lion was deposed as Duke of Saxony and Bavaria by his cousin, Frederick I, Holy Roman Emperor (aka "Barbarossa" for his red beard), Bavaria

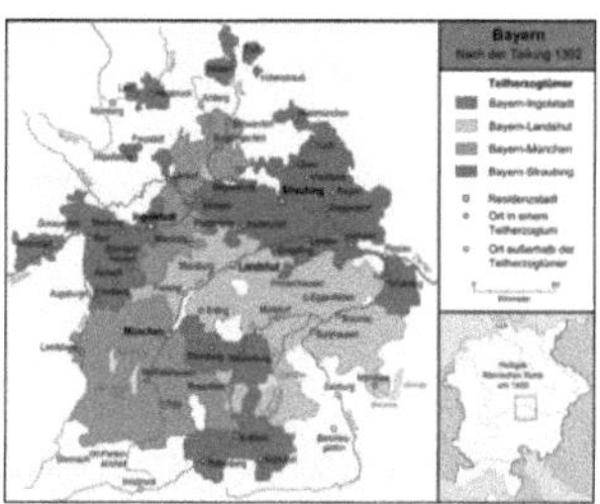

Bavarian duchies after the partition of 1392

was awarded as fief to the Wittelsbach family, counts palatinate of Schyren ("Scheyern" in modern German), which ruled from 1180 to 1918. The Electoral Palatinate by Rhine ("Kurpfalz" in German) was also acquired by the House of Wittelsbach in 1214.

The first of several divisions of the duchy of Bavaria occurred in 1255. With the extinction of the Hohenstaufen in 1268 also Swabian territories were acquired by the Wittelsbach dukes. Emperor Louis the Bavarian acquired Brandenburg, Tirol, Holland and Hainaut for his House but released the Upper Palatinate for the Palatinate branch of the Wittelsbach in 1329. In 1506 with the Landshut War of Succession the other parts of Bavaria were reunited and Munich became the sole capital.

17th and 18th centuries

Further information: Electorate of Bavaria

In 1623 the Bavarian duke replaced his relative, the Count Palatine of the Rhine in the early days of the Thirty Years' War and acquired the powerful prince-electoral dignity in the Holy Roman Empire, determining its Emperor thence forward, as well as special legal status under the empire's laws. The country became one of the Jesuit supported counter-reformation centers. During the early and mid-18th century the ambitions of the Bavarian prince electors led to several wars with Austria as well as occupations by Austria (Spanish succession, election of a Wittelsbach emperor instead of a Habsburger). From 1777 onwards and after the old Bavarian branch of the family had died out with elector Max III Joseph, Bavaria and the Electoral Palatinate were governed once again in personal union, now by the Palatinian lines.

Kingdom of Bavaria

When Napoleon abolished the Holy Roman Empire, Bavaria became a kingdom in 1806, and its area doubled. Tirol was temporarily united, Salzburg temporarily reunited with Bavaria but finally ceded to Austria. In return the Rhenish Palatinate and Franconia were annexed to Bavaria in 1815. Between 1799 and 1817 the leading minister count Montgelas followed a strict policy of modernisation and laid the foundations of administrative structures that survived even the monarchy and are (in their core) valid until today. In 1808 a first and in 1818 a more modern constitution (by the standards of the time) was passed, that established a bicameral Parliament with a House of Lords (*Kammer der Reichsräte*) and a House of Commons (*Kammer der Abgeordneten*). The constitution would last until the collapse of the monarchy at the end of World War I.

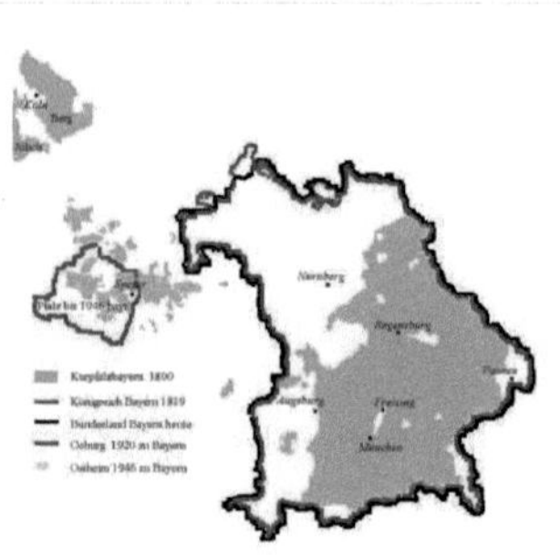

Bavaria in the 19th century and beyond

Bavaria as a part of the German Empire

Bavarian stamps during the German empire period

After the rise of Prussia to prominence Bavaria managed to preserve its independence by playing off the rivalries of Prussia and Austria. Allied to Austria, it was defeated in the 1866 Austro-Prussian War and did not belong to the North German Federation of 1867, but the question of German unity was still alive. When France attacked Prussia in 1870, the south German states Baden, Württemberg and Bavaria joined the Prussian forces and ultimately joined the Federation, which was renamed *Deutsches Reich* (German Empire) in 1871. Bavaria continued as a monarchy, and it even had some special rights within the federation (such as an army, railways and a postal service of its own).

In the early-20th century Wassily Kandinsky, Paul Klee, Henrik Ibsen, and other notable artists were drawn to Bavaria, notably to the Schwabing district of Munich, later devastated by World War II.

20th century

On November 12, 1918, Ludwig III signed a document, the Anif declaration, releasing both civil and military officers from their oaths; the newly-formed republican government of Socialist premier Kurt Eisner interpreted this as an abdication. To date, however, no member of the house of Wittelsbach has ever formally declared renunciation of the throne. On the other hand, none has ever since officially called upon their Bavarian or Stewart claims. Family members are active in cultural and social life, including the head of the house, HRH Duke Franz in Bavaria. They step back from any announcements on public affairs, showing approval or disapproval solely by HRH's presence or absence.

A memorial to soldiers who died in the two World Wars. Village in Bavaria.

Eisner was assassinated in February 1919 ultimately leading to a Communist revolt and the short lived Bavarian Socialist Republic being proclaimed 6 April 1919. After violent suppression by elements of the German Army and notably the Freikorps, the Bavarian Socialist Republic fell in May 1919. The Bamberg Constitution (*Bamberger Verfassung*) was enacted on 12 or 14 August 1919 and came into force on 15 September 1919 creating the Free State of Bavaria within the Weimar Republic. Extremist activity further increased, notably the 1923 Beer Hall Putsch led by the National Socialists, and Munich and Nuremberg became Nazi strongholds under the Third Reich. As a manufacturing center, Munich was heavily bombed during World War II and occupied by U.S. troops. The Rhenish Palatinate was detached from Bavaria in 1946 and made part of the new state Rhineland-Palatinate.

Since World War II, Bavaria has been rehabilitated from a poor agrarian province into a prosperous industrial hub. A massive reconstruction effort restored much of Munich's and other places historic cores. The state capital hosted the 1972 Summer Olympics and matches of the Football World Cups of 1974 and 2006 as well as European Track & Field championships. More recently, former state minister-president Edmund Stoiber was the CDU/CSU candidate for chancellor in the 2002 federal election which he lost, and native son Cardinal Joseph Ratzinger was elected Pope Benedict XVI in 2005.

Coat of arms

The modern coat of arms of Bavaria was designed by Eduard Ege in 1946, following heraldic traditions.

- The Golden Lion: At the dexter chief, sable, a lion rampant Or, armed and langued gules. This represents the administrative region of Upper Palatinate.
- The "Franconian Rake": At the sinister chief, per fess dancetty, gules and argent. This represents the administrative regions of Upper, Middle and Lower Franconia.
- The Blue Panther: At the dexter base, argent, a panther rampant azure, armed Or and langued gules. This represents the regions of Lower and Upper Bavaria.
- The Three Lions: At the sinister base, Or, three lions passant guardant sable, armed and langued gules. This represents Swabia.
- The White-And-Blue inescutcheon: The inescutcheon of white and blue fusils askance was originally the coat of arms of the Counts of Bogen, adopted in 1247 by the Wittelsbachs House. The white-and-blue fusils are indisputably the emblem of Bavaria and these arms today symbolize Bavaria as a whole. Along with the People's Crown, it is officially used as the Minor Coat of Arms.

Bavarian herald Joerg Rugenn wearing a tabard of the arms around 1510

- The People's Crown: The coat of arms is surmounted by a crown with a golden band inset with precious stones and decorated with five ornamental leaves. This crown first appeared in the coat of arms to symbolize sovereignty of the people after the royal crown was eschewed in 1923.

Geography

The Bavarian Alps

Bavaria shares international borders with Austria and the Czech Republic as well as with Switzerland (across Lake Constance). Neighbouring states within Germany are Baden-Württemberg, Hesse, Thuringia and Saxony. Two major rivers flow through the state, the Danube (*Donau*) and the Main. The Bavarian Alps define the border with Austria, (including the Austrian federal-states of Vorarlberg, Tyrol and Salzburg) and within the range is the highest peak in Germany, the Zugspitze. The Bavarian Forest and the Bohemian Forest form the vast majority of the frontier with the Czech Republic and Bohemia.

The major cities in Bavaria are Munich (*München*), Nuremberg (*Nürnberg*), Augsburg, Regensburg, Würzburg, Ingolstadt, Fürth and Erlangen.

Population and area

Administrative region	Population (2010)		Area (km²)		No. municipalities	
Lower Bavaria	1,189,701	9.49%	10,330	14.6%	258	12.5%
Lower Franconia	1,318,695	10.52%	8,531	12.1%	308	15.0%
Upper Franconia	1,071,984	8.55%	7,231	10.2%	214	10.4%
Middle Franconia	1,710,876	13.65%	7,245	10.3%	210	10.2%
Upper Palatinate	1,081,206	8.63%	9,691	13.7%	226	11.0%
Swabia	1,785,875	14.25%	9,992	14.2%	340	16.5%
Upper Bavaria	4,373,588	34.90%	17,530	24.8%	500	24.3%
Total	**12,531,925**	**100.0%**	**70,549**	**100.0%**	**2,056**	**100.0%**

Major cities

City	Inhabitants 31 December 2000	Inhabitants 31 December 2005	Inhabitants 31 December 2008
Munich	1,210,223	1,259,677	1,326,807
Nuremberg	488,400	499,237	503,638
Augsburg	254,982	262,676	263,313
Regensburg	125,676	129,859	133,525
Würzburg	127,966	133,906	133,501
Ingolstadt	115,722	121,314	123,925
Fürth	110,477	113,422	114,071
Erlangen	100,778	103,197	104,980
Bayreuth	74,153	73,997	72,935
Bamberg	69,036	70,081	69,989
Aschaffenburg	67,592	68,642	68,747
Landshut	58,746	61,368	62,606
Kempten (Allgäu)	61,389	61,360	62,135
Rosenheim	58,908	60,226	60,711
Neu-Ulm	50,188	51,410	53,866
Schweinfurt	54,325	54,273	53,588
Passau	50,536	50,651	50,717
Hof	50,741	48,723	47,275
Freising	44,167	45,827	45,654
Straubing	44,014	44,633	44,496

Administrative divisions

Regierungsbezirke (administrative districts)

Bavaria is divided into 7 administrative districts called *Regierungsbezirke* (singular *Regierungsbezirk*).

1. Upper Franconia (German: *Oberfranken*)
2. Middle Franconia (*Mittelfranken*)
3. Lower Franconia (*Unterfranken*)
4. Swabia (*Schwaben*)
5. Upper Palatinate (*Oberpfalz*)
6. Upper Bavaria (*Oberbayern*)
7. Lower Bavaria (*Niederbayern*)

Administrative Districts of Bavaria

Coat of Arms

Upper Bavaria

Lower Bavaria

Upper Palatinate

Upper Franconia

Middle Franconia

Lower Franconia

Swabia

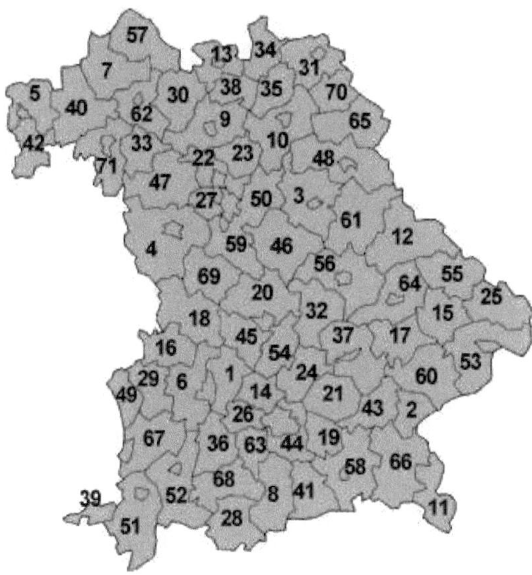

These administrative regions consist of 71 administrative districts (called *Landkreise*, singular *Landkreis*) and 25 independent cities (*kreisfreie Städte*, singular *kreisfreie Stadt*)

Bezirke

Bezirke (districts) are the third communal layer in Bavaria; the others are the Landkreise and the Gemeinden or Städte. In the larger states of Germany (including Bavaria) there are Regierungsbezirke which are only administrative divisions and not self-governing entities as the Bezirke in Bavaria. The Bezirke in Bavaria are territorially identical with the Regierungsbezirke (e.g. Regierung von Oberbayern), but are a different form of administration (having their own parliaments etc.).

Landkreise/kreisfreie cities

Landkreise:

Munich

Nuremberg

Augsburg

Regensburg

Würzburg

Bayreuth

Bamberg

- Aichach-Friedberg
- Altötting
- Amberg-Sulzbach
- Ansbach
- Aschaffenburg
- Augsburg
- Bad Kissingen
- Bad Tölz-Wolfratshausen
- Bamberg
- Bayreuth
- Berchtesgadener Land
- Cham
- Coburg
- Dachau
- Landsberg
- Landshut
- Lichtenfels
- Lindau
- Main-Spessart
- Miesbach
- Miltenberg
- Mühldorf
- München (*Landkreis München*)
- Neuburg-Schrobenhausen
- Neumarkt
- Neustadt (Aisch)-Bad Windsheim
- Neustadt (Waldnaab)
- Neu-Ulm

- Deggendorf

- Nürnberger LandOberallgäuOstallgäuPassau (district)PassauPfaffenhofen (district)PfaffenhofenRegen (district)RegenRegensburg (district)RegensburgRhön-GrabfeldRosenheim (district)RosenheimRoth (district)RothRottal-InnSchwandorf (district)SchwandorfSchweinfurt (district)SchweinfurtStarnberg (district)StarnbergStraubing-BogenTirschenreuth (district)TirschenreuthTraunstein (district)TraunsteinUnterallgäuWeilheim-SchongauWeissenburg-GunzenhausenWunsiedel (district)WunsiedelWürzburg (district)Würzburg

- Dillingen
- Dingolfing-Landau
- Donau-Ries
- Ebersberg
- Eichstätt
- Erding
- Erlangen-Höchstadt
- Forchheim
- Freising
- Freyung-Grafenau
- Fürstenfeldbruck
- Fürth
- Garmisch-Partenkirchen
- Günzburg
- Hassberge
- Hof
- Kelheim
- Kitzingen
- Kronach
- Kulmbach

Kreis-free Cities:

- Amberg
- Ansbach
- Aschaffenburg
- Augsburg
- Bamberg
- Bayreuth
- Coburg
- Erlangen
- Fürth
- Hof
- Ingolstadt
- Kaufbeuren
- Kempten
- Landshut
- Memmingen
- Munich (*München*)
- Nuremberg (*Nürnberg*)
- Passau
- Regensburg
- Rosenheim
- Schwabach
- Schweinfurt
- Straubing
- Weiden
- Würzburg

Gemeinden (municipalities)

The 71 administrative districts are on the lowest level divided into 2031 municipalities (called *Gemeinden*, singular *Gemeinde*). Together with the 25 independent cities (which are in effect municipalities independent of *Landkreis* administrations), there are a total of 2056 municipalities in Bavaria.

In 44 of the 71 administrative districts, there are a total of 215 unincorporated areas (as of January 1, 2005, called *gemeindefreie Gebiete*, singular *gemeindefreies Gebiet*), not belonging to any municipality, all uninhabited, mostly forested areas, but also four lakes (Chiemsee-without islands, Starnberger See-without island Roseninsel, Ammersee, which are the three largest lakes of Bavaria, and Waginger See).

Politics

Bavaria has a multi-party system where the biggest parties are the conservative Christian Social Union of Bavaria (CSU), which has dominated politics since 1945 and won every election since then, and the center-left Social Democratic Party of Germany (SPD). The German green party, Alliance '90/The Greens is represented in the parliament as well. Since 2008 Germany's liberal party, the Free Democratic Party (FDP) and the Free Voters are represented in Bavaria's parliament as well. CSU and FDP agreed in October 2008 to form a coalition, while SPD, Free Voters and the Greens form the opposition.

Bavaria has a unicameral *Landtag*, or state parliament, elected by universal suffrage. Until December 1999, there was also a *Senat*, or Senate, whose members were chosen by social and economic groups in Bavaria, but following a referendum in 1998, this institution was abolished. The head of government is the Minister-President.

In 1995 Bavaria introduced direct democracy on the local level in a referendum. This was initiated bottom-up by an association called *Mehr Demokratie* (More Democracy). This is a grass-roots organization which campaigns for the right to citizen-initiated referendums. In 1997 the Bavarian Supreme Court aggravated the regulations considerably (e.g. by introducing a turn-out quorum). Nevertheless, Bavaria has the most advanced regulations on local direct democracy in Germany. This has led to a spirited citizens' participation in communal and municipal affairs—835 referenda took place from 1995 through 2005.

In the 2003 elections the CSU won more than two thirds of the seats in Landtag—something no party had ever achieved in post-war German history. In the following 2008 elections the CSU lost its absolute majority in the Landtag for the first time in 46 years.[5]

This loss is probably attributed to its push for an anti-smoking law, the most stringent in Germany, which became one of the most controversial laws ever enacted in Bavaria. As result, the CSU changed its stance and weakened the anti-smoking law to allow some more loopholes. However, the citizens held a successful petition drive in November–December 2009 to call for a total smoking ban. The CSU dismissed the petition and demands as unnecessary and frivolous, but the referendum proceeded and the voters voted for the smoking ban.

Minister-presidents of Bavaria since 1945

Minister-presidents of Bavaria					
No.	Name	Born-Died	Party affiliation	Begin of Tenure	End of Tenure
1	Fritz Schäffer	1888–1967	CSU	1945	1945
2	Wilhelm Hoegner	1887–1980	SPD	1945	1946
3	Hans Ehard	1887–1980	CSU	1946	1954
4	Wilhelm Hoegner	1887–1980	SPD	1954	1957
5	Hanns Seidel	1901–1961	CSU	1957	1960
6	Hans Ehard	1887–1980	CSU	1960	1962
7	Alfons Goppel	1905–1991	CSU	1962	1978
8	Franz Josef Strauß	1915–1988	CSU	1978	1988
9	Max Streibl	1932–1998	CSU	1988	1993
10	Edmund Stoiber	*1941	CSU	1993	2007
11	Günther Beckstein	*1943	CSU	2007	2008
12	Horst Seehofer	*1949	CSU	2008	incumbent

German-Bavarian relations

The *Bayernpartei* (Bavaria Party) advocates Bavarian independence from Germany. Bavaria was the only state to reject the West German constitution in 1949, but this did not prevent its implementation. One of Germany's principal political parties, the Christian Democratic Union (CDU), is replaced in Bavaria by the Christian Social Union of Bavaria (CSU), but in practice the two parties cooperate fully in the Bundestag. Bavaria had its own border police force, separate from the Federal Border Guard, until Austria's EU accession in 1995.

Furthermore, the people from the three northern districts of Bavaria known as Franconia (Mittelfranken, Oberfranken and Unterfranken), do not all consider themselves as Bavarian. They have their own history and celebrate their own identity, which is distinct from southern Bavaria, and symbolized by the Franconian rake (Fränkischer Rechen). The flag is often seen during local festivals. Some Franconians would also like to see their own Bundesland Franken "Federal State of Franconia".

Economy

Bavaria has long had one of the largest and healthiest economies of any region in Germany, or Europe for that matter.[6] Its GDP in 2007 exceeded 434 billion Euros (about 600 bn US$).[7] This makes Bavaria itself one of the largest economies in Europe and only 17 countries in the world have higher GDP.[8] Some large companies headquartered in Bavaria include BMW, Siemens, Rohde & Schwarz, Audi, Munich Re, Allianz, Infineon, MAN, Wacker Chemie, Puma AG, and Adidas AG. Bavaria has a GDP per capita of over $48 000 US, meaning that if it were its own independent country it would rank 7th or 8th in the world.

Company names

The motorcycle and automobile makers BMW (*Bayerische Motoren-Werke*, or Bavarian Motor Works) and Audi, Allianz, Grundig (consumer electronics), Siemens (electricity, telephones, informatics, medical instruments), Continental (Automotive Tire and Electronics), Adidas, Puma, HypoVereinsbank (UniCredit Group), Infineon and Krauss-Maffei Wegmann have (or had) a Bavarian industrial base.

Bavaria has also given its name to the largest Colombian brewery (Cervecería Bavaria) and a major Dutch brewery (Biermerk Bavaria).

Culture

Some features of the Bavarian culture and mentality are remarkably distinct from the rest of Germany. Noteworthy differences (especially in rural areas, less significant in the major cities) can be found with respect to:

Bavarian church with Alps in the background

Though only a relatively small part belongs to the Alps, the perception of Bavaria as an alpine region endures.

Religion

While 56.4% of the population adhere to the Catholic Church,[9] 21% are affiliated with the Evangelical Lutheran Church in Bavaria.[10]

Most of Bavaria is predominantly Roman Catholic, but the Evangelical Lutheran Church has a strong presence in large parts of Franconia. Only Saarland has a higher percentage of Catholics among the German states.

The current pope, Benedict XVI (Joseph Alois Ratzinger), was born in Marktl am Inn in Upper Bavaria and was Cardinal-Archbishop of Munich and Freising.

Full-body relic of Saint Hyacinth in the former Cistercian monastery Fürstenfeld Abbey, Bavaria

Traditions

Bavarians commonly emphasize pride in their traditions. Traditional costumes collectively known as Tracht are worn on special occasions and include in Altbayern Lederhosen for males and Dirndl for females. Centuries-old folk music is performed. The Maibaum, or Maypole (which in the Middle Ages served as the community's yellow pages, as figurettes on the pole represent the trades of the village), and the bagpipes in the Upper Palatinate region bear witness to the ancient Celtic and Germanic remnants of cultural heritage of the region.

Whether actually in Bavaria, overseas or full of citizens from other nations they continue to cultivate their traditions. They hold festivals and dances to keep their traditions alive. In New York the German American Cultural Society is a larger umbrella group for others such as the Bavarian organizations, which represent a specific part of Germany. They proudly put forth a German Parade called Steuben Parade each year. Various affiliated events take place amongst its groups, one of which is the Bavarian Dancers.

Food and drink

Bavarians tend to place a great value on food and drink. In addition to their renowned dishes, Bavarians also consume many items of food and drink which are unusual elsewhere in Germany; for example Weisswurst ("white sausage") or in some instances a variety of entrails. At folk festivals and in many beer gardens, beer is traditionally served by the litre (the so-called Maß). Bavarians are particularly proud of the traditional Reinheitsgebot, or purity law, initially established by the Duke of Bavaria for the City of Munich (e.g. the court) in 1487 and the duchy in 1516. According to this law, only three ingredients were allowed in beer: water, barley, and hops. In 1906 the Reinheitsgebot made its way to all-German law, and remained a law in Germany until the EU struck it down recently as incompatible with the European common market. German breweries, however, cling to the principle. Bavarians are also known as some of the world's most beer-loving people with an average annual consumption of 170 litres per person, although figures have been declining in recent years.

Bavaria is also home to the Franconia wine region, which is situated along the Main River in Franconia. The region has produced wine (*Frankenwein*) for over 1,000 years and is famous for its use of the Bocksbeutel wine bottle. The production of wine forms an integral part of the regional culture, and many of its villages and cities hold their own wine festivals (Weinfeste) throughout the year.

Language and dialects

Three German dialects are spoken in Bavaria: Austro-Bavarian in Old Bavaria (South-East and East), Swabian German (an Alemannic German dialect) in the Bavarian part of Swabia (South West) and East Franconian German in Franconia (North).

Ethnography

Bavarians consider themselves to be egalitarian and informal. Their sociability can be experienced at the annual Oktoberfest, the world's largest beer festival, which welcomes around six million visitors every year, or in the famous beer gardens. In traditional Bavarian beer gardens, patrons may bring their own food but only buy beer from the brewery that runs the beer garden.

In the United States, particularly among German Americans, Bavarian culture is viewed somewhat nostalgically, and many "Bavarian

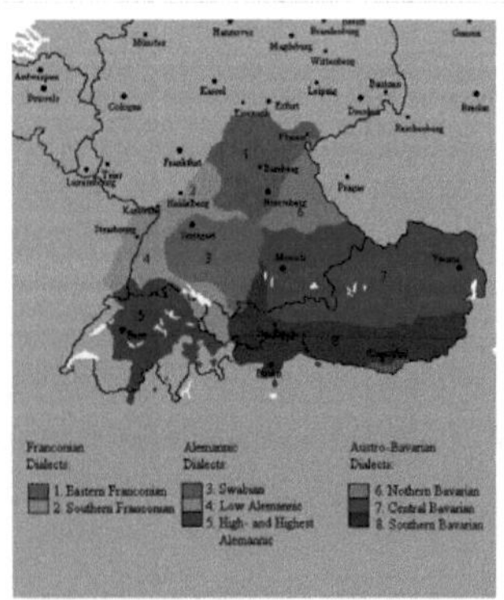

Upper German, southern counterpart to Central German, both forming the High German Languages

villages", most notably Frankenmuth, Michigan and Leavenworth, Washington, have been founded. Since 1962, the latter has been styled with a Bavarian theme; it is also home to "one of the world's largest collections of nutcrackers" and an Oktoberfest celebration it claims is among the most attended in the world outside of Munich.[11]

Historical buildings

Johannisburg Castle in
Aschaffenburg

Würzburg Residence

Fortress Marienberg and the Alte
Mainbrücke in Würzburg

Plassenburg Castle in
Kulmbach

Cathedral in
Bamberg

Basilica of the
Vierzehnheiligen

Castle of Coburg

Festspielhaus of Richard Wagner
in Bayreuth

Imperial Castle in Nuremberg

Kastell Biriciana, Weissenburg
close to the Limes

Kreuztor in
Ingolstadt

Castle of Neuburg an der Donau

Old Stone Bridge and
Cathedral of Regensburg

Walhalla temple in Donaustauf
near Regensburg

Befreiungshalle in Kelheim

Cathedral and Oberhaus
fortification in Passau

Trausnitz castle, Landshut

Burghausen Castle

Townhall in Augsburg

Frauenkirche in Munich

Residenz in Munich

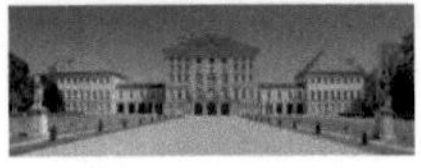

Nymphenburg Palace in Munich

Cathedral in Freising

Herrenchiemsee Palace

Linderhof Palace

Hohenschwangau Castle

Wieskirche, Steingaden

Church of St. Bartholomew at Königssee

Neuschwanstein

Neuschwanstein Castle

Neuschwanstein was built for King Ludwig II, as a second home. It remains unfinished.

Famous people

There are many famous people who were born or lived in present-day Bavaria:

- **Popes** Pope Benedict XVI (baptismal name: **Joseph Aloisius Ratzinger**)—the current Pope of the Roman Catholic Church; Pope Damasus II and Pope Victor II.
- **Painters** such as Hans Holbein the Elder, Albrecht Dürer, Albrecht Altdorfer, Lucas Cranach, Carl Spitzweg, Franz von Lenbach, Franz von Stuck, Franz Marc, Paul Klee, Erwin Eisch, Gabriele Münter.
- **Musicians** such as Johannes Heesters, Orlando di Lasso, Christoph Willibald Gluck, Richard Wagner (originally from Saxony), Richard Strauss, Carl Orff, Johann Pachelbel and Theobald Boehm, the inventor of the modern flute, and countertenor Klaus Nomi.
- Modern musicians Klaus Doldinger, Barbara Dennerlein, Hans-Jürgen Buchner.
- Opera singers like Jonas Kaufmann and Diana Damrau.
- **Writers**, **poets** and **playwrights** like Hans Sachs, Jean Paul, Frank Wedekind, Christian Morgenstern, Oskar Maria Graf, Bertolt Brecht, Lion Feuchtwanger, Thomas Mann and his sons Klaus and Golo Mann, Karl Marx lived in Munich for a few years, Ludwig Thoma.

- **Scientists** such as Max Planck, Wilhelm Conrad Röntgen, and Werner Heisenberg, as well as Adam Ries, Joseph von Fraunhofer, Georg Ohm, Johannes Stark, Carl von Linde, Ludwig Prandtl, Rudolf Moessbauer, Lothar Rohde and Hermann Schwarz, Helmut Hirt and Robert Huber.
- Well-known **inventors** such as Martin Behaim, Levi Strauss and Rudolf Diesel.
- **Physicians** like Max Joseph von Pettenkofer, Sebastian Kneipp and the neurologist Alois Alzheimer, who first described Alzheimer's Disease.
- **Football players** like Max Morlock, Karl Mai, Franz Beckenbauer, Sepp Maier, Gerd Müller, Paul Breitner, Bernd Schuster, Klaus Augenthaler, Lothar Matthäus, Philipp Lahm, Bastian Schweinsteiger, Holger Badstuber, Thomas Müller, Dietmar Hamann and Stefan Reuter
- **Other sportspeople** such as golfer Bernhard Langer and basketball player Dirk Nowitzki
- **Actors** like Werner Stocker, Helmut Fischer, Walter Sedlmayr, Gustl Bayrhammer, Ottfried Fischer, Ruth Drexel, Elmar Wepper, Fritz Wepper, Uschi Glas, Yank Azman, Wilfried Klaus(Soko 312).
- **Film directors** Rainer Werner Fassbinder, Joseph Vilsmaier, Werner Herzog, Franz Xaver Bogner.
- **Mysterious people**: Kaspar Hauser (the famous foundling), The Smith of Kochel (legend).
- **Legendary outlaws** such as Mathias Kneißl, the legendary robber or Matthias Klostermayr, better known as Bavarian Hiasl
- Noted automobile designer Peter Schreyer, born in Bad Reichenhall
- Fictional characters such as X-Men's Nightcrawler (comics)

See also

- Outline of Germany
- List of rulers of Bavaria
- List of Premiers of Bavaria
- Former countries in Europe after 1815
- Extensive pictures of Bavaria in addition to those shown below are linked from in Category:Bavaria, where they are organized (predominantly) by locale.
- Template:User Bavarian Ancestry Bavarian ancestry infobox

References

[1] "State population" (http://www.statistik-portal.de/Statistik-Portal/de_zs01_by.asp). *Portal of the Federal Statistics Office Germany.* August 2011. . Retrieved 2011-04-17.
[2] "State GDP" (http://www.statistik-portal.de/Statistik-Portal/en/en_jb27_jahrtab65.asp). *Portal of the Federal Statistics Office Germany.* . Retrieved 2010-05-03.
[3] http://www.bayern.de/English-.594.htm
[4] The Germanic realms in pre-Carolingian Central Europe, Herbert Schutz, 2000, p. 313
[5] n-tv:Fiasko für die CSU (http://www.n-tv.de/Koalitionspartner_noetig_Fiasko_fuer_die_CSU/280920085818/1030029.html)
[6] Its GDP is 143% of the EU average (as of 2005) against a German average of 121.5%, see Eurostat (http://epp.eurostat.ec.europa.eu/extraction/evalight/EVAlight.jsp?A=1&language=en&root=/theme1/reg/reg_e2gdp)
[7] Gemeinsames Datenangebot der Statistischen Ämter des Bundes und der Länder (http://www.statistik.baden-wuerttemberg.de/Statistik-Portal/de_jb27_jahrtab65.asp)
[8] See the list of countries by GDP (nominal).
[9] chiesa cattolica http://www.dbk.de/imperia/md/content/kirchlichestatistik/bev-kath-l__nd-2008.pdf
[10] EKD http://www.ekd.de/download/kirchenmitglieder_2007.pdf
[11] Leavenworth, Washington The Bavarian Village (http://www.leavenworth.org/)

External links

- Churches (http://commons.wikimedia.org/wiki/Churches_in_Bavaria)
- Official Tourism Board (http://www.bavaria.by)
- Official government website (http://www.bayern.de)
- Platform with additional information about many cities in Bavaria (http://www.bayregio.de)
- Tradition and Culture (http://www.bavariandancers.com)
- Foreign Trade (http://www.auwi-bayern.de)
- statistics (http://www.statistik.bayern.de/veroeffentlichungen/)

Germany

<table>
<tr><td colspan="3" align="center">Federal Republic of Germany
Bundesrepublik Deutschland</td></tr>
<tr><td colspan="3">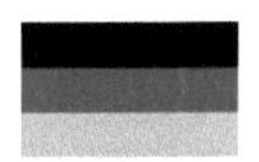 </td></tr>
<tr><td colspan="3" align="center">Anthem:
The third stanza of Das Lied der Deutschen
The Song of the Germans</td></tr>
<tr><td colspan="3" align="center">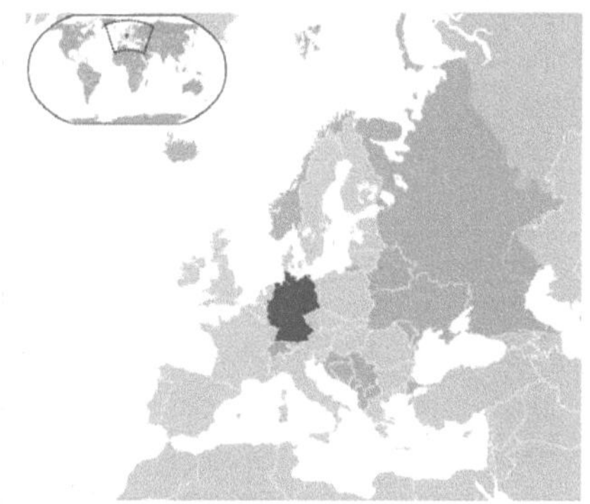
Location of Germany(dark green)

— in Europe(green & dark grey)
— in the European Union(green) — [Legend]</td></tr>
<tr><td colspan="2">Capital
(and largest city)</td><td>Berlin
52°31′N 13°23′E</td></tr>
<tr><td colspan="2" align="center">Official language(s)</td><td>German[1]</td></tr>
<tr><td colspan="2" align="center">Demonym</td><td>German</td></tr>
<tr><td colspan="2" align="center">Government</td><td>Federal parliamentary constitutional republic</td></tr>
<tr><td>-</td><td>President</td><td>Christian Wulff (CDU membership dormant)</td></tr>
<tr><td>-</td><td>Chancellor</td><td>Angela Merkel (CDU)</td></tr>
<tr><td>-</td><td>President of the Bundestag</td><td>Norbert Lammert (CDU)</td></tr>
<tr><td>-</td><td>President of the Bundesrat</td><td>Horst Seehofer (CSU)</td></tr>
<tr><td colspan="3" align="center">Formation</td></tr>
<tr><td>-</td><td>Holy Roman Empire</td><td>2 February 962</td></tr>
<tr><td>-</td><td>Unification</td><td>18 January 1871</td></tr>
<tr><td>-</td><td>Federal Republic</td><td>23 May 1949</td></tr>
</table>

-	Reunification	3 October 1990
Area		
-	Total	357,021 km^2 (63rd) 137,847 sq mi
-	Water (%)	2.416
Population		
-	2010 estimate	81,799,600[1] (15th)
-	Density	229/km^2 (55th) 593/sq mi
GDP (PPP)		2011 estimate
-	Total	$3.089 trillion[2] (5th)
-	Per capita	$37,935[2] (18th)
GDP (nominal)		2011 estimate
-	Total	$3.628 trillion[2] (4th)
-	Per capita	$44,555[2] (19th)
Gini (2006)		27 (low)
HDI (2011)		▲ 0.905[3] (very high) (9th)
Currency		Euro (€)^2(2002 – present) (EUR)
Time zone		CET (UTC+1)
-	Summer (DST)	CEST (UTC+2)
Drives on the		right
ISO 3166 code		DE
Internet TLD		.de ^{3}
Calling code		49
1	Danish, Low German, Sorbian, Romany and Frisian are officially recognised by the ECRML.	
2	Before 2002: Deutsche Mark (DEM).	
3	Also .eu, shared with European Union member states.	

Germany (🔊 [ˈdʒɜrməni/), officially the **Federal Republic of Germany** (German: *Bundesrepublik Deutschland*, pronounced [ˈbʊndəsʁepuˌbliːk ˈdɔʏtʃlant] (🔊 listen)),[4] is a federal parliamentary republic in Europe. The country consists of 16 states while the capital and largest city is Berlin. Germany covers an area of 357,021 km^2 and has a largely temperate seasonal climate. With 81.8 million inhabitants, it is the most populous member state and the largest economy in the European Union. It is one of the major political powers of the European continent and a technological leader in many fields.

A region named Germania, inhabited by several Germanic peoples, was documented before AD 100. During the Migration Age, the Germanic tribes expanded southward, and established successor kingdoms throughout much of Europe. Beginning in the 10th century, German territories formed a central part of the Holy Roman Empire.[5] During the 16th century, northern German regions became the centre of the Protestant Reformation while southern and western parts remained dominated by Roman Catholic denominations, with the two factions clashing in the

Thirty Years' War, marking the beginning of the Catholic–Protestant divide that has characterized German society ever since.[6] Occupied during the Napoleonic Wars, the rise of Pan-Germanism inside the German Confederation resulted in the unification of most of the German states into the German Empire in 1871 which was Prussian dominated. After the German Revolution of 1918–1919 and the subsequent military surrender in World War I, the Empire was replaced by the Weimar Republic in 1918, and partitioned in the Versailles Treaty. Amidst the Great Depression, the Third Reich was proclaimed in 1933. The latter period was marked by Fascism and the Second World War. After 1945, Germany was divided by allied occupation, and evolved into two states, East Germany and West Germany. In 1990 Germany was reunified.

Germany was a founding member of the European Community in 1957, which became the EU in 1993. It is part of the Schengen Area and since 1999 a member of the eurozone. Germany is a member of the United Nations, NATO, the G8, the G20, the OECD and the Council of Europe, and took a non-permanent seat on the UN Security Council for the 2011–2012 term.

It has the world's fourth largest economy by nominal GDP and the fifth largest by purchasing power parity. It is the second largest exporter and third largest importer of goods. The country has developed a very high standard of living and a comprehensive system of social security. Germany has been the home of many influential scientists and inventors, and is known for its cultural and political history.

Etymology

The English word *Germany* derives from the Latin Germania, which came into use after Julius Caesar adopted it for the peoples east of the Rhine.[7] In other languages it has various names.

The German term *Deutschland* (originally *diutisciu land*, "the German lands") is derived from *deutsch*, descended from Old High German *diutisc* "popular" (i. e., belonging to the *diot* or *diota* "people"; originally used to distinguish the language of the common people from Latin and its Romance descendants). This in turn descends from Proto-Germanic **þiudiskaz* "popular" (see also the Latinised form Theodiscus), derived from **þeudō*, descended from Proto-Indo-European **tewtéh₂-* "people".[8]

History

Germanic tribes and Frankish Empire

Map of the Germania and the Roman Empire

The Germanic tribes are thought to date from the Nordic Bronze Age or the Pre-Roman Iron Age. From southern Scandinavia and north Germany, they expanded south, east and west from the 1st century BC, coming into contact with the Celtic tribes of Gaul as well as Iranian, Baltic, and Slavic tribes in Eastern Europe.[9] Under Augustus, the Roman General Publius Quinctilius Varus began to invade Germania (an area extending roughly from the Rhine to the Ural Mountains). In AD 9, three Roman legions led by Varus were defeated by the Cheruscan leader Arminius. By AD 100, when Tacitus wrote *Germania*, Germanic tribes had settled along the Rhine and the Danube (the Limes Germanicus), occupying most of the area of modern Germany; Austria, southern Bavaria and the western Rhineland, however, were Roman provinces.[10]

In the 3rd century a number of large West Germanic tribes emerged: Alamanni, Franks, Chatti, Saxons, Frisii, Sicambri, and Thuringii. Around 260, the Germanic peoples broke into Roman-controlled lands.[11] After the invasion of the Huns in 375, and with the decline of Rome from 395, Germanic tribes moved further south-west. Simultaneously several large tribes formed in what is now Germany and displaced the smaller Germanic tribes. Large areas (known since the Merovingian period as Austrasia) were occupied by the Franks, and Northern Germany was ruled by the Saxons and Slavs.[10]

Holy Roman Empire

Martin Luther initiated the Protestant Reformation.

On 25 December 800, Charlemagne founded the Carolingian Empire, which was divided in 843.[12] The Holy Roman Empire resulted from the eastern portion of this division. Its territory stretched from the Eider River in the north to the Mediterranean coast in the south.[12] Under the reign of the Ottonian emperors (919–1024), several major duchies were consolidated, and the German king was crowned Holy Roman Emperor of these regions in 962. The Holy Roman Empire absorbed northern Italy and Burgundy under the reign of the Salian emperors (1024–1125), although the emperors lost power through the Investiture Controversy.

Under the Hohenstaufen emperors (1138–1254), the German princes increased their influence further south and east into territories inhabited by Slavs, preceding German settlement in these areas and further east *(Ostsiedlung)*. Northern German towns grew prosperous as members of the Hanseatic League.[13] Starting with the Great Famine in 1315, then the Black Death of 1348–50, the population of Germany plummeted.[14] The edict of the Golden Bull in 1356 provided the basic constitution of the empire and codified the election of the emperor by seven prince-electors who ruled some of the most powerful principalities and archbishoprics.[15]

Martin Luther publicised his 95 Theses in 1517, challenging the Roman Catholic Church and initiating the Protestant Reformation. A separate Lutheran church became the official religion in many German states after 1530. Religious conflict led to the Thirty Years' War (1618–1648), which devastated German lands.[16] The population of the German states was reduced by about 30%.[17] The Peace of Westphalia (1648) ended religious warfare among the German states, but the empire was *de facto* divided into numerous independent principalities. From 1740 onwards, dualism between the Austrian Habsburg Monarchy and the Kingdom of Prussia dominated German history. In 1806, the *Imperium* was overrun and dissolved as a result of the Napoleonic Wars.[18]

German Confederation and Empire

Following the fall of Napoleon I of France, the Congress of Vienna convened in 1814 and founded the German Confederation (Deutscher Bund), a loose league of 39 sovereign states. Disagreement with restoration politics partly led to the rise of liberal movements, followed by new measures of repression by Austrian statesman Metternich. The *Zollverein*, a tariff union, furthered economic unity in the German states.[19] National and liberal ideals of the French Revolution gained increasing support among many, especially young, Germans. In the light of a series of revolutionary movements in Europe, which established a republic in France, intellectuals and commoners started the Revolutions of 1848 in the German states. King Frederick William IV of Prussia was offered the title of Emperor, but with a loss of power; he rejected the crown and the proposed constitution, leading to a temporary setback for the movement.[20]

Conflict between King William I of Prussia and the increasingly liberal parliament erupted over military reforms in 1862, and the king appointed Otto von Bismarck the new Prime Minister of Prussia. Bismarck successfully waged war on Denmark in 1864. Prussian victory in the Austro-Prussian War of 1866 enabled him to create the North German Federation (Norddeutscher Bund) and to exclude Austria, formerly the leading German state, from the federation's affairs. After the French defeat in the Franco-Prussian War, the German Empire was proclaimed 1871 in Versailles, uniting all scattered parts of Germany except Austria (*Kleindeutschland*, or "Lesser Germany"). With almost two thirds of its territory and population, Prussia was the dominating constituent of the new state;

Foundation of the German Empire in Versailles, 1871. Bismarck is at the centre in a white uniform.

the Hohenzollern King of Prussia ruled as its concurrent Emperor, and Berlin became its capital.[20] In the *Gründerzeit* period following the unification of Germany, Bismarck's foreign policy as Chancellor of Germany under Emperor William I secured Germany's position as a great nation by forging alliances, isolating France by diplomatic means, and avoiding war. Under Wilhelm II, however, Germany, like other European powers, took an imperialistic course leading to friction with neighbouring countries. As a result of the Berlin Conference in 1884 Germany claimed several colonies including German East Africa, German South-West Africa, Togo, and Cameroon.[21] Most alliances in which Germany had previously been involved were not renewed, and new alliances excluded the country.[22]

The assassination of Austria's crown prince on 28 June 1914 triggered World War I. Germany, as part of the Central Powers, suffered defeat against the Allies in one of the bloodiest conflicts of all time. An estimated two million German soldiers died in World War I.[23] The German Revolution broke out in November 1918, and Emperor Wilhelm II and all German ruling princes abdicated. An armistice ended the war on 11 November, and Germany was forced to sign the Treaty of Versailles in June 1919. The treaty was perceived in Germany as a humiliating continuation of the war, and is often cited as an influence in the rise of Nazism.[24]

Weimar Republic and Third Reich

Adolf Hitler, chancellor and president[1] 1933–19451: office formally vacant from August 1934; Hitler styled himself *"Führer und Reichskanzler"*[25]

At the beginning of the German Revolution in November 1918, Germany was declared a republic. However, the struggle for power continued, with radical-left communists seizing power in Bavaria. The revolution came to an end on 11 August 1919, when the Weimar Constitution was signed by President Friedrich Ebert.[26] Suffering from the Great Depression, the harsh peace conditions dictated by the Treaty of Versailles, and a long succession of unstable governments, Germans increasingly lacked identification with the government. This was exacerbated by a widespread right-wing *Dolchstoßlegende*, or stab-in-the-back myth, which argued that Germany had lost World War I because of those who wanted to overthrow the government. The Weimar government was accused of betraying Germany by signing the Versailles Treaty. By 1932, the German Communist Party and the Nazi Party controlled the majority of parliament, fuelled by discontent with the Weimar government. After a series of unsuccessful cabinets, President Paul von Hindenburg appointed Adolf Hitler as Chancellor of Germany on 30 January 1933.[27] On 27 February 1933 the Reichstag building went up in flames, and a consequent emergency decree abrogated basic citizens' rights. An Enabling Act passed in parliament gave Hitler unrestricted legislative power. Only the Social Democratic Party voted against it, while Communist MPs had already been imprisoned.[28] [29] Using his powers to crush any actual or potential resistance, Hitler established a centralised totalitarian state within months. Industry was revitalised with a focus on military rearmament.[30]

In 1935, Germany reacquired control of the Saar and in 1936 military control of the Rhineland, both of which had been lost in the Treaty of Versailles.[31] In 1938 and 1939, Austria and Czechoslovakia were brought under German control and the invasion of Poland was prepared through the Molotov–Ribbentrop pact and Operation Himmler. On 1 September 1939 the German Wehrmacht launched a blitzkrieg on Poland, which was swiftly occupied by Germany and by the Soviet Red Army. The UK and France declared war on Germany, marking the beginning of World War II.[32] As the war progressed, Germany and its allies quickly gained control of much of continental Europe though plans to occupy the United Kingdom failed. On 22 June 1941, Germany broke the Molotov–Ribbentrop pact and invaded the Soviet Union. Japan's attack on Pearl Harbor led Germany to declare war on the United States. The Battle of Stalingrad forced the German army to retreat on the Eastern front.[32] In September 1943, Germany's ally Italy surrendered, and German troops were forced to defend an additional front in Italy. D-Day opened a Western front, as Allied forces advanced towards German territory. On 8 May 1945, the German armed forces surrendered after the Red Army occupied Berlin.[33]

In what later became known as The Holocaust, the Third Reich regime had enacted policies directly subjugating many dissidents and minorities. Millions of people were murdered by the Nazis during the Holocaust, including a sizeable number of Jews, Gypsies, Jehovah's Witnesses, Poles and other Slavs, including Soviet POWs, people with mental and/or physical disabilities, homosexuals, and members of the political opposition.[34] World War II was responsible for more than 40 million dead in Europe.[35] The Nuremberg trials of Nazi war criminals were held after World War II.[36] The war casualties for Germany are estimated at 5.3 million German soldiers[37] millions of

Berlin in ruins after World War II

German civilians;[38] [39] [40] [41] [42] and losing the war resulted in large territorial losses; the expulsion of about 15 million Germans from the eastern areas of Germany and other countries; mass rape of German women;[43] and the destruction of multiple major cities.

East and West Germany

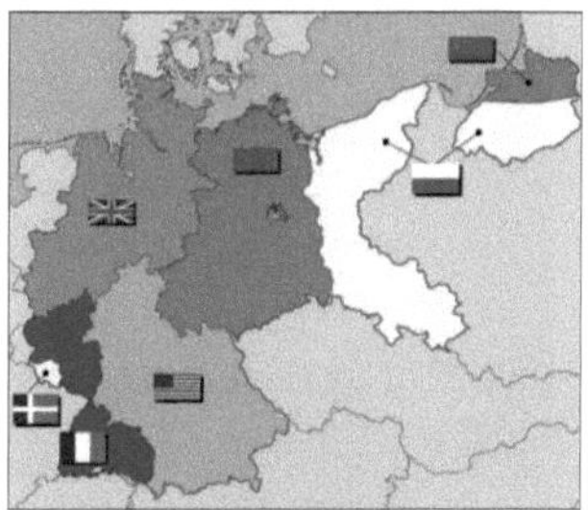

Occupation zones in Germany, 1947. The territories east of the Oder-Neisse line, under Polish and Soviet *de jure* administration and *de facto* annexation, are shown as white as is the detached Saar protectorate.

After the surrender of Germany, the remaining German territory and Berlin were partitioned by the Allies into four military occupation zones. The western sectors, controlled by France, the United Kingdom, and the United States, were merged on 23 May 1949 to form the *Federal Republic of Germany (Bundesrepublik Deutschland)*; on 7 October 1949, the Soviet Zone became the *German Democratic Republic (Deutsche Demokratische Republik*, or DDR). They were informally known as "West Germany" and "East Germany". East Germany selected East Berlin as its capital, while West Germany chose Bonn as a provisional capital, to emphasise its stance that the two-state solution was an artificial and temporary *status quo*.[44]

West Germany, established as a federal parliamentary republic with a "social market economy", was allied with the United States, the UK and France. The country enjoyed prolonged economic growth beginning in the early 1950s (*Wirtschaftswunder*). West Germany joined NATO in 1955 and was a founding member of the European Economic Community in 1957. East Germany was an Eastern bloc state under political and military control by the USSR via the latter's occupation forces and the Warsaw Pact. Though East Germany claimed to be a democracy, political power was exercised solely by leading members (*Politbüro*) of the communist-controlled Socialist Unity Party of Germany (SED), supported by the Stasi, an immense secret service,[45] and a variety of sub-organisations controlling every aspect of society. A Soviet-style command economy was set up; the GDR later became a Comecon state.[46] While East German propaganda was based on the benefits of the GDR's social programmes and the alleged constant threat of a West German invasion, many of her citizens looked to the West for freedom and prosperity.[47] The Berlin Wall, built in 1961 to stop East Germans from escaping to West Germany, became a symbol of the Cold War.[20]

Tensions between East and West Germany were reduced in the early 1970s by Chancellor Willy Brandt's *Ostpolitik*. In summer 1989, Hungary decided to dismantle the Iron Curtain and open the borders, causing the emigration of thousands of East Germans to West Germany via Hungary. This had devastating effects on the GDR, where regular mass demonstrations received increasing support. The East German authorities unexpectedly eased the border restrictions, allowing East German citizens to travel to the West; originally intended to help retain East Germany as a state, the opening of the border actually led to an acceleration of the *Wende* reform process. This culminated in the *Two Plus Four Treaty* a year later on 12 September 1990, under which the four occupying powers renounced

The Berlin Wall in front of the Brandenburg Gate shortly before its fall in 1989

their rights under the Instrument of Surrender, and Germany regained full sovereignty. This permitted German reunification on 3 October 1990, with the accession of the five re-established states of the former GDR (new states or "neue Länder").[20]

Berlin Republic and the EU

Based on the Berlin/Bonn Act, adopted on 10 March 1994, Berlin once again became the capital of the reunified Germany, while Bonn obtained the unique status of a *Bundesstadt* (federal city) retaining some federal ministries.[48] The relocation of the government was completed in 1999.[49] Since reunification, Germany has taken a more active role in the European Union and NATO. Germany sent a peacekeeping force to secure stability in the Balkans and sent a force of German troops to Afghanistan as part of a NATO effort to provide security in that country after the ousting of the Taliban.[50] These deployments were controversial since, after the war, Germany was bound by domestic law only to deploy troops for defence roles.[51] In 2005, Angela Merkel became the first female Chancellor of Germany as the leader of a grand coalition.[20]

Geography

Germany is in Western and Central Europe, bordering Denmark in the north, Poland and the Czech Republic in the east, Austria and Switzerland in the south, France and Luxembourg in the south-west, and Belgium and the Netherlands in the north-west. It lies mostly between latitudes 47° and 55° N (the tip of Sylt is just north of 55°), and longitudes 5° and 16° E. The territory covers 357021 km^2 (sq mi), consisting of 349223 km^2 (sq mi) of land and 7798 km^2 (3011 sq mi) of water. It is the seventh largest country by area in Europe and the 62nd largest in the world.[52]

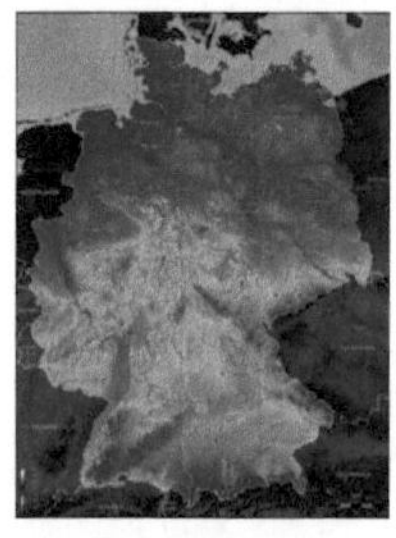
Topographic map

Elevation ranges from the mountains of the Alps (highest point: the Zugspitze at 2962 metres / 9718 feet) in the south to the shores of the North Sea (Nordsee) in the north-west and the Baltic Sea (Ostsee) in the north-east. The forested uplands of central Germany and the lowlands of northern Germany (lowest point: Wilstermarsch at 3.54 metres / 11.6 feet below sea level) are traversed by such major rivers as the Rhine, Danube and Elbe. Glaciers are found in the Alpine region, but are experiencing deglaciation. Significant natural resources are iron ore, coal, potash, timber, lignite, uranium, copper, natural gas, salt, nickel, arable land and water.[52]

Climate

Most of Germany has a temperate seasonal climate in which humid westerly winds predominate. The climate is moderated by the North Atlantic Drift, the northern extension of the Gulf Stream. This warmer water affects the areas bordering the North Sea; consequently in the north-west and the north the climate is oceanic. Rainfall occurs year-round, especially in the summer. Winters are mild and summers tend to be cool, though temperatures can exceed 30 °C (86 °F).[53]

The east has a more continental climate; winters can be very cold and summers very warm, and long dry periods are frequent. Central and southern Germany are transition regions which vary from moderately oceanic to continental. In addition to the maritime and continental climates that predominate over most of the country, the Alpine regions in the extreme south and, to a lesser degree, some areas of the Central German Uplands have a mountain climate, characterised by lower temperatures and greater precipitation.[53]

Biodiversity

The territory of Germany can be subdivided into two ecoregions: European-Mediterranean montane mixed forests and Northeast-Atlantic shelf marine.[54] As of 2008 the majority of Germany is covered by either arable land (34%) or forest and woodland (30.1%); only 13.4% of the area consists of permanent pastures, 11.8% is covered by settlements and streets.[55]

The eagle is a protected bird of prey and the national heraldic animal.

Plants and animals are those generally common to middle Europe. Beeches, oaks, and other deciduous trees constitute one third of the forests; conifers are increasing as a result of reforestation. Spruce and fir trees predominate in the upper mountains, while pine and larch are found in sandy soil. There are many species of ferns, flowers, fungi, and mosses. Wild animals include deer, wild boar, mouflon, fox, badger, hare, and small numbers of beavers.[56]

The national parks in Germany include the Wadden Sea National Parks, the Jasmund National Park, the Vorpommern Lagoon Area National Park, the Müritz National Park, the Lower Oder Valley National Park, the Harz National Park, the Saxon Switzerland National Park and the Bavarian Forest National Park. More than 400 registered zoos and animal parks operate in Germany, which is believed to be the largest number in any country.[57] The Zoologische Garten Berlin is the oldest zoo in Germany and presents the most comprehensive collection of species in the world.[58]

Politics

The Reichstag in Berlin is the site of the German parliament (Bundestag).

Germany is a federal, parliamentary, representative democratic republic. The German political system operates under a framework laid out in the 1949 constitutional document known as the *Grundgesetz* (Basic Law). Amendments generally require a two-thirds majority of both chambers of parliament; the fundamental principles of the constitution, as expressed in the articles guaranteeing human dignity, the separation of powers, the federal structure, and the rule of law are valid in perpetuity.[59]

The president, currently Christian Wulff, is the head of state and invested primarily with representative responsibilities and powers. He is elected by the *Bundesversammlung* (federal convention), an institution consisting of the members of the *Bundestag* and an equal number of state delegates. The second-highest official in the German order of precedence is the *Bundestagspräsident* (President of the *Bundestag*), who is elected by the *Bundestag* and responsible for overseeing the daily sessions of the body. The third-highest official and the head of government is the Chancellor, who is appointed by the *Bundespräsident* after being elected by the *Bundestag*.[20]

The chancellor, currently Angela Merkel, is the head of government and exercises executive power, similar to the role of a Prime Minister in other parliamentary democracies. Federal legislative power is vested in the parliament consisting of the *Bundestag* (Federal Diet) and *Bundesrat* (Federal Council), which together form the legislative body. The *Bundestag* is elected through direct elections, by proportional representation (mixed-member).[52] The members of the *Bundesrat* represent the governments of the sixteen federated states and are members of the state cabinets.[20]

Since 1949, the party system has been dominated by the Christian Democratic Union and the Social Democratic Party of Germany with all chancellors hitherto being member of either party. However, the smaller liberal Free Democratic Party (which has had members in the *Bundestag* since 1949) and the Alliance '90/The Greens (which has controlled seats in parliament since 1983) have also played important roles.[60]

Germany has a civil law system based on Roman law with some references to Germanic law. The *Bundesverfassungsgericht* (Federal Constitutional Court) is the German Supreme Court responsible for constitutional matters, with power of judicial review.[20] [61] Germany's supreme court system, called *Oberste Gerichtshöfe des Bundes*, is specialised: for civil and criminal cases, the highest court of appeal is the inquisitorial Federal Court of Justice, and for other affairs the courts are the Federal Labour Court, the Federal Social Court, the Federal Finance Court and the Federal Administrative Court. The *Völkerstrafgesetzbuch* regulates the consequences of crimes against humanity, genocide and war crimes, and gives German courts universal jurisdiction in some circumstances.[62] Criminal and private laws are codified on the national level in the *Strafgesetzbuch* and the *Bürgerliches Gesetzbuch* respectively. The German penal system is aimed towards rehabilitation of the criminal and the protection of the general public.[63] Except for petty crimes, which are tried before a single professional judge, and serious political crimes, all charges are tried before mixed tribunals on which lay judges (*Schöffen*) sit side by side with professional judges.[64] [65]

Constituent states

Germany comprises sixteen states that are collectively referred to as *Länder*.[66] Each state has its own state constitution[67] and is largely autonomous in regard to its internal organisation. Due to differences in size and population the subdivision of these states varies, especially between city states (*Stadtstaaten*) and states with larger territories (*Flächenländer*). For regional administrative purposes five states, namely Baden-Württemberg, Bavaria, Hesse, North Rhine-Westphalia and Saxony, consist of a total of 22 Government Districts (*Regierungsbezirke*). As of 2009 Germany is divided into 403 districts (*Kreise*) on municipal level, these consist of 301 rural districts and 102 urban districts.[68]

Lower Saxony

Bremen

Hamburg

Mecklenburg-Vorpommern

Saxony-Anhalt

Saxony

Brandenburg

Berlin

Thuringia

Hesse

North Rhine-Westphalia

Rhineland-Palatinate

Bavaria

Baden-Württemberg

Saarland

Schleswig-Holstein

State	Capital	Area (km²)	Population
Baden-Württemberg	Stuttgart	35,752	10,717,000
Bavaria	Munich	70,549	12,444,000
Berlin	Berlin	892	3,400,000
Brandenburg	Potsdam	29,477	2,568,000
Bremen	Bremen	404	663,000
Hamburg	Hamburg	755	1,735,000
Hesse	Wiesbaden	21,115	6,098,000
Mecklenburg-Vorpommern	Schwerin	23,174	1,720,000
Lower Saxony	Hanover	47,618	8,001,000
North Rhine-Westphalia	Düsseldorf	34,043	18,075,000
Rhineland-Palatinate	Mainz	19,847	4,061,000
Saarland	Saarbrücken	2,569	1,056,000
Saxony	Dresden	18,416	4,296,000
Saxony-Anhalt	Magdeburg	20,445	2,494,000
Schleswig-Holstein	Kiel	15,763	2,829,000
Thuringia	Erfurt	16,172	2,355,000

Foreign relations

Germany has a network of 229 diplomatic missions abroad and maintains relations with more than 190 countries.[69] As of 2011 it is the largest contributor to the budget of the European Union (providing 20%)[70] and the third largest contributor to the UN (providing 8%).[71] Germany is a member of NATO, the Organisation of Economic Co-operation and Development (OECD), the G8, the G20, the World Bank and the International Monetary Fund (IMF). It has played a leading role in the European Union since its inception and has maintained a strong alliance with France since the end of World War II. Germany seeks to advance the creation of a more unified European political, defence, and security apparatus.[72] [73]

Chancellor Angela Merkel hosting the G8 summit in Heiligendamm

The development policy of the Federal Republic of Germany is an independent area of German foreign policy. It is formulated by the Federal Ministry for Economic Cooperation and Development (BMZ) and carried out by the implementing organisations. The German government sees development policy as a joint responsibility of the international community.[74] It is the world's third biggest aid donor after the United States and France.[75] [76]

During the Cold War, Germany's partition by the Iron Curtain made it a symbol of East-West tensions and a political battleground in Europe. However, Willy Brandt's Ostpolitik was a key factor in the *détente* of the 1970s.[77] In 1999, Chancellor Gerhard Schröder's government defined a new basis for German foreign policy by taking part in the NATO decisions surrounding the Kosovo War and by sending German troops into combat for the first time since World War II.[78] The governments of Germany and the United States are close political allies.[20] The 1948 Marshall Plan and strong cultural ties have crafted a strong bond between the two countries, although Schröder's vocal opposition to the Iraq War suggested the end of Atlanticism and a relative cooling of German-American relations.[79] The two countries are also economically interdependent: 8.8% of German exports are U.S.-bound and 6.6% of German imports originate from the U.S.[80]

Military

Germany's military, the *Bundeswehr*, is organized in *Heer* (Army), *Marine* (Navy), *Luftwaffe* (Air Force), *Zentraler Sanitätsdienst* (Central Medical Services) and *Streitkräftebasis* (Joint Support Service) branches. As of 2005, military spending was an estimated 1.5% of the country's GDP, that is position 99 in a ranking of all countries;[52] absolutely, German military expenditure is the eighth-highest in the world.[81] In peacetime, the Bundeswehr is commanded by the Minister of Defence. If Germany went to war, which according to the constitution is allowed only for defensive purposes, the Chancellor would become commander in chief of the *Bundeswehr*.[82]

As of May 2011 the Bundeswehr employs 188,000 professional soldiers, 31,000 18–25 year-old conscripts who serve for at least six months.[83] The German government plans to reduce the number of soldiers to 170,000 professionals and up to 15,000 short-time volunteers (*voluntary military service*).[84] Reservists are available to the Armed Forces and participate in defence exercises and deployments abroad, a new reserve concept of their future strength and functions was announced 2011.[84] As of April 2011, the German military had about 6,900 troops stationed in foreign countries as part of international peacekeeping forces, including about 4,900 Bundeswehr troops in the NATO-led ISAF force in Afghanistan and Uzbekistan, 1,150 German soldiers in Kosovo, and 300 troops with UNIFIL in Lebanon.[85]

Until 2011, military service was compulsory for men at age 18, and conscripts served six-month tours of duty; conscientious objectors could instead opt for an equal length of *Zivildienst* (civilian service), or a six-year commitment to (voluntary) emergency services like a fire department or the Red Cross. On 1 July 2011 conscription was officially suspended and replaced with a voluntary service.[86] [87] Since 2001 women may serve in all functions of service without restriction, but they are not subject to conscription. There are presently some 17,500 women on active duty and a number of female reservists.[88]

Economy

Germany has a social market economy with a highly qualified labour force, a large capital stock, a low level of corruption,[90] and a high level of innovation.[91] It has the largest national economy in Europe, the fourth largest by nominal GDP in the world,[92] and the fifth largest by PPP[92] in 2009. The service sector contributes approximately 71% of the total GDP, industry 28%, and agriculture 0.9%.[52] The average national unemployment rate in 2010 was about 7.5%.[52] First estimates indicate a 3.6% increase in the price-adjusted GDP for 2010, following a 4.7% drop in 2009.[93]

A Mercedes-Benz car. Germany was the world's leading exporter of goods from 2003 to 2008.[89]

Germany is a founding member of the EU, the G8 and the G20, and was the world's largest exporter from 2003 to 2008. In 2009 it remained the second largest exporter and third largest importer of goods. Most of the country's exports are in engineering, especially machinery, automobiles, chemical goods and metals.[52] Germany is a leading producer of wind turbines and solar-power technology.[94] Annual trade fairs and congresses are held in cities throughout Germany.[95]

Germany is an advocate of closer European economic and political integration. Its commercial policies are increasingly determined by agreements among European Union (EU) members and by EU legislation. Germany introduced the common European currency, the euro, on 1 January 2002.[96] [97] Its monetary policy is set by the European Central Bank. Two decades after German reunification, standards of living and per capita incomes remain significantly higher in the states of the former West Germany than in the former East.[98] The modernisation and integration of the eastern German economy is a long-term process scheduled to last until the year 2019, with annual transfers from west to east amounting to roughly $80 billion.[99] In January 2009 the German government approved a €50 billion economic stimulus plan to protect several sectors from a downturn and a subsequent rise in unemployment rates.[100]

Of the world's 500 largest stock-market-listed companies measured by revenue in 2010, the Fortune Global 500, 37 are headquartered in Germany. 30 Germany-based companies are included in the DAX, the German stock market index. Well-known global brands are Mercedes-Benz, BMW, SAP, Siemens, Volkswagen, Adidas, Audi, Allianz, Porsche, and Nivea.[101] Germany is recognised for its specialised small and medium enterprises. Around 1,000 of these companies are global market leaders in their segment and are labelled hidden champions.[102]

The list includes the largest companies by turnover in 2009. Unranked are the largest bank and the largest insurance company in 2007:

Rank[103]	Name	Headquarters	Revenue (Mil. €)	Profit (Mil. €)	Employees (World)
1	Volkswagen AG	Wolfsburg	108,897	4,120	329,305
2	Daimler AG	Stuttgart	99,399	3,985	272,382
3	Siemens AG	Munich/Berlin	72,488	3,806	398,200
4	E.ON AG	Düsseldorf	68,731	7,204	87,815
5	Metro AG	Düsseldorf	64,337	825	242,378
6	Deutsche Post AG	Bonn	63,512	1,389	475,100
7	Deutsche Telekom AG	Bonn	62,516	569	241,426
8	BASF SE	Ludwigshafen	57,951	4,065	95,175
9	BMW AG	Munich	56,018	3,126	107,539
10	ThyssenKrupp AG	Essen/Duisburg	51,723	2,102	191,350

Infrastructure

The ICE 3 train

With its central position in Europe, Germany is a transport hub. This is reflected in its dense and modern transport networks. The motorway (Autobahn) network ranks as the third largest worldwide in length.[104] Germany has established a polycentric network of high-speed trains. The InterCityExpress or *ICE* network of the Deutsche Bahn serves major German cities as well as destinations in neighbouring countries.[105] The largest German airports are Frankfurt Airport and Munich Airport, both hubs of Lufthansa, while Air Berlin has hubs at Berlin Tegel and Düsseldorf. Other major airports include Berlin Schönefeld, Hamburg, Cologne/Bonn and Leipzig/Halle. Both airports in Berlin will be consolidated at a site adjacent to Berlin Schönefeld, which will become Berlin Brandenburg Airport in 2012.[106]

As of 2008, Germany was the world's sixth largest consumer of energy,[107] and 60% of its primary energy was imported.[108] Government policy promotes energy conservation and renewable energy. Energy efficiency has been improving since the early 1970s; the government aims to meet the country's electricity demands using only renewable sources by 2050.[109] In 2010, energy sources were: oil (33.7%); coal, including lignite (22.9%); natural gas (21.8%); nuclear (10.8%); hydro-electric and wind power (1.5%); and other renewable sources (7.9%).[110] In 2000, the government and the nuclear power industry agreed to phase out all nuclear power plants by 2021.[111] Germany is committed to the Kyoto protocol and several other treaties promoting biodiversity, low emission standards, recycling, and the use of renewable energy, and supports sustainable development at a global level.[112] The German government has initiated wide-ranging emission reduction activities and the country's overall emissions are falling.[113] Nevertheless the country's greenhouse gas emissions were the highest in the EU as of 2007.[114]

Science and technology

Germany's achievements in sciences have been significant, and research and development efforts form an integral part of the economy.[115] The Nobel Prize has been awarded to 103 German laureates.[116] For most of the 20th century, German laureates had more awards than those of any other nation, especially in the sciences (physics, chemistry, and physiology or medicine).[117] [118]

The work of Albert Einstein and Max Planck was crucial to the foundation of modern physics, which Werner Heisenberg and Max Born developed further.[119] Wilhelm Conrad Röntgen discovered X-rays and was the first winner of the Nobel Prize in Physics in 1901.[120] Numerous mathematicians were born in Germany, including Carl Friedrich Gauss, David Hilbert, Bernhard Riemann, Gottfried Leibniz, Karl Weierstrass, Hermann Weyl and Felix Klein. Research institutions in Germany include the Max Planck Society, the Helmholtz Association and the Fraunhofer Society. The Gottfried Wilhelm Leibniz Prize is granted to ten scientists and academics every year. With a maximum of €2.5 million per award it is one of highest endowed research prizes in the world.[121]

Albert Einstein

Germany has been the home of many famous inventors and engineers, such as Johannes Gutenberg, credited with the invention of movable type printing in Europe; Hans Geiger, the creator of the Geiger counter; and Konrad Zuse, who built the first fully automatic digital computer.[122] German inventors, engineers and industrialists such as Count Ferdinand von Zeppelin, Otto Lilienthal, Gottlieb Daimler, Rudolf Diesel, Hugo Junkers and Karl Benz helped shape modern automotive and air transportation technology.[123] Aerospace engineer Wernher von Braun developed the first space rocket and later on was a prominent member of NASA and developed the Saturn V Moon rocket, which paved the way for the success of the US Apollo program. Heinrich Rudolf Hertz's work in the domain of electromagnetic radiation was pivotal to the development of modern telecommunication.[124]

Germany is also one of the leading countries in developing and using green technologies. Companies specializing in green technology have an estimated turnover of 200€ billion. Especially the expertise in engineering, science and research of Germany is eminently respectable. The lead markets of Germany's green technology industry are power generation, sustainable mobility, material efficiency, energy efficiency, waste management and recycling, sustainable water management.[125]

Demographics

With its estimated population of 81.8 million in January 2010,[1] Germany is the most populous country in the European Union and ranks as the 15th most populous country in the world.[126] Its population density stands at 229.4 inhabitants per square kilometre. The overall life expectancy in Germany at birth is 79.9 years. The fertility rate of 1.4 children per mother, or 7.9 births per 1000 inhabitants in 2009, is one of the lowest in the world.[127] Since the 1990s, Germany's death rate has continuously exceeded its birth rate.[128] The Federal Statistical Office of Germany forecast that the population will shrink to between 65 and 70 million by 2060 (depending on the level of net migration).[129]

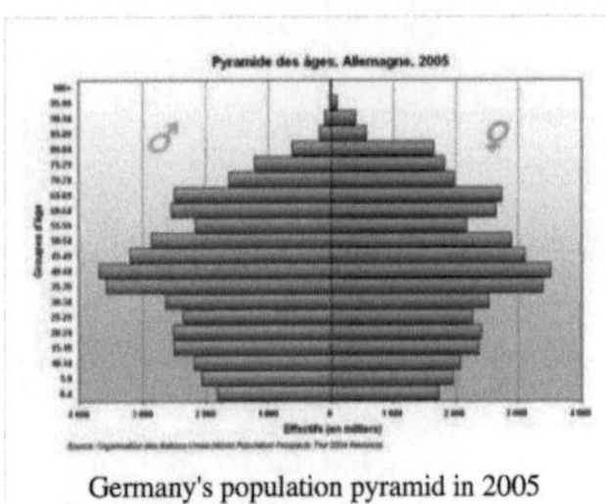

Germany's population pyramid in 2005

German nationals make up 91% of the population of Germany. As of 2009, about seven million foreign citizens were registered in Germany, and 19% of the country's residents were of foreign or partially foreign descent (including persons descending or partially descending from ethnic German repatriates), 96% of whom lived in Western Germany or Berlin.[130] The United Nations Population Fund lists Germany as host to the third-highest number of international migrants worldwide, about 5% or 10 million of all 191 million migrants.[131] As a consequence of restrictions to Germany's formerly rather unrestricted laws on asylum and immigration, the number of immigrants seeking asylum or claiming German ethnicity (mostly from the former Soviet Union) has been declining steadily since 2000.[132] In 2009, 20% of the population had immigrant roots, the highest since 1945.[133] As of 2008, the largest national group was from Turkey (2.5 million), followed by Italy (776,000) and Poland (687,000).[134] About 3 million "Aussiedler"—ethnic Germans, mainly from Eastern Europe and the former Soviet Union—have resettled in Germany since 1987.[135]

Germany has a number of large cities. The largest conurbation is the Rhine-Ruhr region (11.5 million as of 2006), including Düsseldorf (the capital of North Rhine-Westphalia), Cologne, Dortmund, Essen, Duisburg, and Bochum.[136]

Religion

Christianity is the largest religion in Germany, with around 51.5 million adherents (62.8%) in 2008,[137] of which 30.0% are Catholics and 29.9% are Protestants, belonging to the Evangelical Church in Germany (EKD); the remainder consists of small denominations (each less than 0.5% of the German population).[138] Protestantism is concentrated in the north and east and Roman Catholicism is concentrated in the south and west;[6] 1.6% of the country's overall population declare themselves Orthodox Christians.[137]

The Cologne Cathedral at the Rhine river is a UNESCO World Heritage Site.

The second largest religion is Islam with an estimated 3.8 to 4.3 million adherents (4.6% to 5.2%),[139] followed by Buddhism with 250,000 and Judaism with around 200,000 adherents (0.3%); Hinduism has some 90,000 adherents (0.1%). All other religious communities in Germany have fewer than 50,000 adherents.[140] Of the roughly 4 million Muslims, most are Sunnis and Alevites from Turkey, but there are a small number of Shi'ites and other denominations.[139] German Muslims, a large portion of whom are of Turkish origin, lack full official state recognition of their religious community.[6] Germany has Europe's third largest Jewish population (after France and the United Kingdom).[141] Approximately 50% of the Buddhists in Germany are Asian immigrants.[142]

Germans with no stated religious adherence make up 34.1% of the population, especially in the former East Germany and major metropolitan areas.[138] German reunification in 1990 greatly increased the country's non-religious population, a legacy of the state atheism of the previously Soviet-controlled East. Christian church membership has decreased in recent decades, particularly among Protestants.[6]

Languages

German is the official and predominant spoken language in Germany.[143] It is one of 23 official languages in the European Union, and one of the three working languages of the European Commission. Recognised native minority languages in Germany are Danish, Low German, Sorbian, Romany, and Frisian; they are officially protected by the ECRML. The most used immigrant languages are Turkish, Kurdish, Polish, the Balkan languages, and Russian; 67% of German citizens claim to be able to communicate in at least one foreign language and 27% in at least two languages other than their own.[143]

Standard German is a West Germanic language and is closely related to and classified alongside English, Low German, Dutch, and the Frisian languages. To a lesser extent, it is also related to the East (extinct) and North Germanic languages. Most German vocabulary is derived from the Germanic branch of the Indo-European language family.[144] Significant minorities of words are derived from Latin and Greek, with a smaller amount from French and most recently English (known as Denglisch). German is written using the Latin alphabet. German dialects, traditional local varieties traced back to the Germanic tribes, are distinguished from varieties of standard German by their lexicon, phonology, and syntax.[145]

Education

The University of Heidelberg was established in 1386.

Over 99% of Germans age 15 and above are estimated to be able to read and write.[52] However, a growing number of inhabitants are functionally illiterate.[146] Responsibility for educational oversight in Germany lies primarily with the individual federated states. Since the 1960s, a reform movement attempted to unify secondary education in a *Gesamtschule* (comprehensive school); several West German states later simplified their school system to two or three tiers. A system of apprenticeship called *Duale Ausbildung* ("dual education") allows pupils in vocational training to learn in a company as well as in a state-run vocational school.[147]

Optional kindergarten education is provided for all children between three and six years old, after which school attendance is compulsory for at least nine years. Primary education usually lasts for four years and public schools are not stratified at this stage.[147] In contrast, secondary education includes three traditional types of schools focused on different levels of academic ability: the *Gymnasium* enrols the most gifted children and prepares students for university studies; the *Realschule* for intermediate students lasts six years; the *Hauptschule* prepares pupils for vocational education.[148]

The general entrance requirement for university is Abitur, a qualification normally based on continuous assessment during the last few years at school and final examinations; however there are a number of exceptions, and precise requirements vary, depending on the state, the university and the subject. Germany's universities are recognised internationally; in the Academic Ranking of World Universities (ARWU) for 2008, six of the top 100 universities in the world are in Germany, and 18 of the top 200.[149] Nearly all German universities are public institutions, charging tuition fees of €50–500 per semester for each student.[150]

Health

Germany has the world's oldest universal health care system, dating back to Otto von Bismarck's Social legislation in 1883.[151] Currently the population is covered by a basic health insurance plan provided by statute. According to the World Health Organization, Germany's health care system was 77% government-funded and 23% privately funded as of 2005.[152] In 2005, Germany spent 11% of its GDP on health care. Germany ranked 20th in the world in life expectancy with 77 years for men and 82 years for women, and it had a very low infant mortality rate (4 per 1,000 live births).[152]

As of 2009, the principal cause of death was cardiovascular disease, at 42%, followed by malignant tumours, at 25%.[153] As of 2008, about 82,000 Germans had been infected with HIV/AIDS and 26,000 had died from the disease (cumulatively, since 1982).[154] According to a 2005 survey, 27% of German adults are smokers.[154] A 2007 study shows Germany has the highest number of overweight people in Europe.[155] [156]

Culture

From its roots, culture in Germany has been shaped by major intellectual and popular currents in Europe, both religious and secular. Historically Germany has been called *Das Land der Dichter und Denker* (the land of poets and thinkers).[157] The federated states are in charge of the cultural institutions. There are 240 subsidised theatres, hundreds of symphonic orchestras, thousands of museums and over 25,000 libraries spread in Germany. These cultural opportunities are enjoyed by many: there are over 91 million German museum visits every year; annually, 20 million go to theatres and operas; 3.6 million per year listen to the symphonic orchestras.[158] The UNESCO inscribed 33 properties in Germany on the World Heritage List.[159]

Ludwig van Beethoven (1770–1827), composer

Germany has established a high level of gender equality,[160] promotes disability rights, and is legally and socially tolerant towards homosexuals. Gays and lesbians can legally adopt their partner's biological children, and civil unions have been permitted since 2001.[161] Germany has also changed its attitude towards immigrants; since the mid-1990s, the government and the majority of Germans have begun to acknowledge that controlled immigration should be allowed based on qualification standards.[162] Germany has been named the world's second most valued nation among 50 countries in 2010.[163] A global opinion poll for the BBC revealed that Germany is recognised for having the most positive influence in the world in 2011.[164]

Arts

J.S.Bach	L.v. Beethoven	R. Wagner
Toccata und Fuge	Symphonie 5 c-moll	Die Walküre

Numerous German painters have enjoyed international prestige through their work in diverse artistic styles. Hans Holbein the Younger, Matthias Grünewald, and Albrecht Dürer were important artists of the Renaissance, Caspar David Friedrich of Romanticism, and Max Ernst of Surrealism. Architectural contributions from Germany include the Carolingian and Ottonian styles, which were precursors of Romanesque. The region later became the site of Gothic, Renaissance and Baroque art. Germany was particularly important in the early modern movement, especially through the Bauhaus movement founded by Walter Gropius. Ludwig Mies van der Rohe became one of the world's most renowned architects in the second half of the 20th century. He conceived of the glass façade skyscraper.[165]

German music includes works by some of the world's most well-known classical music composers, including Ludwig van Beethoven, Johann Sebastian Bach, Johannes Brahms, and Richard Wagner. As of 2008, Germany is the fourth largest music market in the world[166] and has influenced popular music through artists such as Kraftwerk, Alphaville, Boney M., Nena, Nico, Nina Hagen, Scorpions, Die Toten Hosen, Tokio Hotel, Rammstein, and Paul van Dyk.

Literature and philosophy

German literature can be traced back to the Middle Ages and the works of writers such as Walther von der Vogelweide and Wolfram von Eschenbach. Well-known German authors include Johann Wolfgang von Goethe and Friedrich Schiller. The collections of folk tales published by the Brothers Grimm popularised German folklore on an international level. Influential authors of the 20th century include Thomas Mann, Bertolt Brecht, Hermann Hesse, Heinrich Böll, and Günter Grass.[167] German-speaking book publishers produce some 700 million books every year, with about 80,000 titles, nearly 60,000 of them new. Germany comes third in quantity of books published, after the English-speaking book market and the People's Republic of China.[168] The Frankfurt Book Fair is the most important in the world for international deals and trading, with a tradition spanning over 500 years.[169]

The Brothers Grimm

German philosophy is historically significant. Gottfried Leibniz's contributions to rationalism; the establishment of classical German idealism by Immanuel Kant, Johann Gottlieb Fichte, Georg Wilhelm Friedrich Hegel and Friedrich Wilhelm Joseph Schelling; Arthur Schopenhauer's composition of metaphysical pessimism; the formulation of communist theory by Karl Marx and Friedrich Engels; Friedrich Nietzsche's development of perspectivism; Gottlob Frege's contributions to the dawn of analytic philosophy; Martin Heidegger's works on Being; and the development of the Frankfurt school by Max Horkheimer, Theodor Adorno, Herbert Marcuse and Jürgen Habermas have been particularly influential. In the 21st century Germany has contributed to the development of contemporary analytic philosophy in continental Europe, along with France, Austria, Switzerland and the Scandinavian countries.[170]

Media

German cinema dates back to the earliest years of the medium with the work of Max Skladanowsky, which was particularly influential with German expressionists such as Robert Wiene and Friedrich Wilhelm Murnau. Director Fritz Lang's *Metropolis* (1927) is referred to as the first modern science-fiction film. In 1930 the Austrian-American Josef von Sternberg directed *The Blue Angel*, the first major German sound film.[171] During the 1970s and 1980s, New German Cinema directors such as Volker Schlöndorff, Werner Herzog, Wim Wenders, and Rainer Werner Fassbinder put West German cinema on the international stage.[172] The annual European Film Awards ceremony is held every other year in Berlin, home of the European Film Academy (EFA); the Berlin Film Festival, held annually since 1951, is one of the world's foremost film festivals.[173]

More recently, films such as *Good Bye Lenin!* (2003), *Gegen die Wand (Head-on)* (2004), *Der Untergang (Downfall)* (2004), and *Der Baader Meinhof Komplex* (2008) have had international success. The Academy Award for Best Foreign Language Film went to the German production *Die Blechtrommel (The Tin Drum)* in 1979, to *Nowhere in Africa* in 2002, and to *Das Leben der Anderen (The Lives of Others)* in 2007.[174] Germany's television market is the largest in Europe, with some 34 million TV households. Around 90% of German households have cable or satellite TV, with a variety of free-to-view public and commercial channels.[175]

Cuisine

German cuisine varies from region to region. The southern regions of Bavaria and Swabia, for instance, share a culinary culture with Switzerland and Austria. In all regions, meat is often eaten in sausage form.[176] Organic food has gained a market share of ca. 2%, and is expected to increase further.[177] Although wine is becoming more popular in many parts of Germany, the national alcoholic drink is beer. German beer consumption per person is declining, but at 116 litres annually it is still among the highest in the world.[178] The Michelin guide has awarded nine restaurants in Germany three stars, the highest designation, while 15 more received two stars.[179] German restaurants have become the world's second-most decorated after France.[180]

A *Schwarzwälder Kirschtorte* (literally, "Black Forest cherry torte".)

Sports

Signal Iduna Park is the biggest stadium in Germany

Twenty-seven million Germans are members of a sports club and an additional twelve million pursue sports individually.[181] Association football is the most popular sport. With more than 6.3 million official members, the German Football Association (*Deutscher Fußball-Bund*) is the largest sports organisation of its kind worldwide.[181] The Bundesliga attracts the second highest average attendance of any professional sports league in the world.

The German national football team won the FIFA World Cup in 1954, 1974 and 1990 and the UEFA European Football Championship in 1972, 1980 and 1996. Germany hosted the FIFA World Cup in 1974 and 2006 and the UEFA European Football Championship in 1988. Among the most well-known footballers are Franz Beckenbauer, Gerd Müller, Jürgen Klinsmann, Lothar Matthäus, and Oliver Kahn. Other popular spectator sports include handball, volleyball, basketball, ice hockey, and tennis.[181]

Germany is one of the leading motor sports countries in the world. Constructors like BMW and Mercedes are prominent manufacturers in motor sport. Additionally, Porsche has won the 24 Hours of Le Mans, an annual endurance race held in France, 16 times, and Audi has won it 9 times. Formula One driver Michael Schumacher has set many motor sport records during his career, having won more Formula One World Drivers' Championships and more Formula One races than any other driver; he is one of the highest paid sportsmen in history.[182]

Historically, German sportsmen have been successful contenders in the Olympic Games, ranking third in an all-time Olympic Games medal count, combining East and West German medals. In the 2008 Summer Olympics, Germany finished fifth in the medal count,[183] while in the 2006 Winter Olympics they finished first.[184] Germany has hosted the Summer Olympic Games twice, in Berlin in 1936 and in Munich in 1972. The Winter Olympic Games took place in Germany once in 1936 in the twin towns of Garmisch and Partenkirchen.

References

[1] *Key Figures on Europe* (http://epp.eurostat.ec.europa.eu/cache/ITY_OFFPUB/KS-EI-11-001/EN/KS-EI-11-001-EN.PDF). Belgium: European Union. 2011. p. 37. doi:10.2785/623. ISBN 978-92-79-18441-3. .

[2] "World Economic Outlook Database" (http://www.imf.org/external/pubs/ft/weo/2011/02/weodata/weorept.aspx?pr.x=41&pr.y=5& sy=2009&ey=2016&scsm=1&ssd=1&sort=country&ds=.&br=1&c=134&s=NGDPD,NGDPDPC,PPPGDP,PPPPC,LP&grp=0&a=). International Monetary Fund. September 2011. . Retrieved 17-November-2011.

[3] "Human development index" (http://hdr.undp.org/en/media/HDR_2011_EN_Table1.pdf). United Nations Development Programme. 2011. . Retrieved 5 November 2011.

[4] Mangold, Max, ed (1995) (in German). *Duden, Aussprachewörterbuch* (6th ed.). Dudenverlag. pp. 271, 53f. ISBN 9783411209163.

[5] The Latin name *Sacrum Imperium* (Holy Empire) is documented as far back as 1157. The Latin name *Sacrum Romanum Imperium* (Holy Roman Empire) was first documented in 1254. The full name "Holy Roman Empire of the German Nation" (*Heiliges Römisches Reich Deutscher Nation*) dates back to the 15th century.
Zippelius, Reinhold (2006) [1994] (in German). *Kleine deutsche Verfassungsgeschichte: vom frühen Mittelalter bis zur Gegenwart [Brief German Constitutional History: from the Early Middle Ages to the Present]* (7 ed.). Munich: Beck. p. 25. ISBN 9783406476389.

[6] "Germany" (http://berkleycenter.georgetown.edu/resources/countries/germany). Berkley Center for Religion, Peace, and World Affairs. . Retrieved 2011-12-15.

[7] Schulze, Hagen (1998). *Germany: A New History*. Harvard University Press. p. 4. ISBN 0-674-80688-3.

[8] Lloyd, Albert L.; Lühr, Rosemarie; Springer, Otto (1998) (in German). *Etymologisches Wörterbuch des Althochdeutschen, Band II* (http://books.google.com/books?id=iKfYGNwwNVIC&pg=PA523). Göttingen: Vandenhoeck & Ruprecht. pp. 699–704. ISBN 3-525-20768-9. . (for *diutisc*) Lloyd, Albert L.; Lühr, Rosemarie; Springer, Otto (1998) (in German). *Etymologisches Wörterbuch des Althochdeutschen, Band II* (http://books.google.com/books?id=iKfYGNwwNVIC&pg=PA516). Göttingen: Vandenhoeck & Ruprecht. pp. 685–686. ISBN 3-525-20768-9. . (for *diot*)

[9] Claster, Jill N. (1982). *Medieval Experience: 300–1400*. New York University Press. p. 35. ISBN 0-8147-1381-5.

[10] Fulbrook 1991, pp. 9–13.

[11] Bowman, Alan K.; Garnsey, Peter; Cameron, Averil (2005). *The crisis of empire, A.D. 193–337*. The Cambridge Ancient History. **12**. Cambridge University Press. p. 442. ISBN 0-521-30199-8.

[12] Fulbrook 1991, p. 11.

[13] Fulbrook 1991, pp. 13–24.

[14] Nelson, Lynn Harry. *The Great Famine (1315–1317) and the Black Death (1346–1351)* (http://www.vlib.us/medieval/lectures/black_death.html). University of Kansas. . Retrieved 19 March 2011.

[15] Fulbrook 1991, p. 27.

[16] Philpott, Daniel (January 2000). "The Religious Roots of Modern International Relations". *World Politics* **52** (2): 206–245.

[17] Macfarlane, Alan (1997). *The savage wars of peace: England, Japan and the Malthusian trap*. Blackwell. p. 51. ISBN 9780631181170.

[18] Fulbrook 1991, p. 97.

[19] Henderson, W. O. (January 1934). "The Zollverein". *History* **19** (73): 1–19. doi:10.1111/j.1468-229X.1934.tb01791.x.

[20] "Germany" (http://www.state.gov/r/pa/ei/bgn/3997.htm). U.S. Department of State. 10 November 2010. . Retrieved 26 March 2011.

[21] Black, John, ed (2005). *100 maps*. Sterling Publishing. p. 202. ISBN 9781402728853.

[22] Fulbrook 1991, pp. 135, 149.

[23] Crossland, David (22 January 2008). "Last German World War I Veteran Believed to Have Died" (http://www.spiegel.de/international/germany/0,1518,530319,00.html). *Spiegel Online*. . Retrieved 25 March 2011.

[24] Lee, Stephen J. (2003). *Europe, 1890–1945*. Routledge. p. 131. ISBN 9780415254557.

[25] Thamer, Hans-Ulrich (2003). "Beginn der nationalsozialistischen Herrschaft (Teil 2)" (http://www.bpb.de/publikationen/02735619745887542775109928829773,5,0,Beginn_der_nationalsozialistischen_Herrschaft_(Teil_2).html#art5) (in german). *Nationalsozialismus I*. Bonn: Federal Agency for Civic Education. . Retrieved 4 October 2011. "President von Hindenburg died on August 2nd, 1934. The day before, the cabinet had approved a submission making Hitler his successor. The office of the president was to be dissolved and united with that of the chancellor under the name „Führer und Reichskanzler". However, this was in breach of the Enabling Act. (shortened & paraphrased)"

[26] Fulbrook 1991, pp. 156–160.

[27] Fulbrook 1991, pp. 155–158, 172–177.

[28] "Das Ermächtigungsgesetz 1933" (http://www.dhm.de/lemo/html/nazi/innenpolitik/ermaechtigungsgesetz/index.html) (in German). Deutsches Historisches Museum. . Retrieved 25 March 2011.

[29] Stackelberg, Roderick (1999). *Hitler's Germany: Origins, interpretations, legacies*. Routledge. p. 103. ISBN 9780415201155.

[30] "Industrie und Wirtschaft" (http://www.dhm.de/lemo/html/nazi/wirtschaft/index.html) (in German). Deutsches Historisches Museum. . Retrieved 25 March 2011.

[31] Fulbrook 1991, pp. 188–189.

[32] Fulbrook 1991, pp. 190–195.

[33] Steinberg, Heinz Günter (1991) (in German). *Die Bevölkerungsentwicklung in Deutschland im Zweiten Weltkrieg: mit einem Überblick über die Entwicklung von 1945 bis 1990*. Kulturstiftung der dt. Vertriebenen. ISBN 9783885570899.

[34] Niewyk, Donald L.; Nicosia, Francis R. (2000). *The Columbia Guide to the Holocaust*. Columbia University Press. pp. 45–52. ISBN 9780231112000.

[35] "Leaders mourn Soviet wartime dead" (http://news.bbc.co.uk/2/hi/europe/4530565.stm). *BBC News*. 9 May 2005. . Retrieved 18 March 2011.

[36] Overy, Richard (17 February 2011). "Nuremberg: Nazis on Trial" (http://www.bbc.co.uk/history/worldwars/wwtwo/ nuremberg_article_01.shtml). BBC History. . Retrieved 25 March 2011.

[37] Rüdiger Overmans. *Deutsche militärische Verluste im Zweiten Weltkrieg*. Oldenbourg 2000. ISBN 3-486-56531-1

[38] Das Deutsche Reich und der Zweite Weltkrieg, Bd. 9/1, ISBN 3-421-06236-6. Page 460 (This study was prepared by the German Armed Forces Military History Research Office, an agency of the German government)

[39] Peter Antill; Peter Dennis (10 October 2005). *Berlin 1945: end of the Thousand Year Reich* (http://books.google.com/ books?id=vAzgsCDUky0C). Osprey Publishing. p. 85. ISBN 9781841769158. . Retrieved 24 June 2011.

[40] Bonn : Kulturstiftung der Deutschen Vertriebenen, *Vertreibung und Vertreibungsverbrechen, 1945–1948 : Bericht des Bundesarchivs vom 28. Mai 1974 : Archivalien und ausgewählte Erlebnisberichte* / [Redaktion, Silke Spieler]. Bonn :1989 ISBN 388557067X. (This is a study of German expulsion casualties due to "war crimes" prepared by the German government Archives)

[41] Germany reports. With an introd. by Konrad Adenauer. Germany (West). Presse- und Informationsamt. Wiesbaden, Distribution: F. Steiner, 1961] Page 32

[42] Robert N. Proctor, *Racial Hygiene: Medicine under the Nazis*, Harvard 1988,

[43] Beevor, Antony (2003) [2002]. *Berlin: The downfall 1945*. Penguin. pp. 31–32, 409–412. ISBN 9780140286960.

[44] Wise, Michael Z. (1998). *Capital dilemma: Germany's search for a new architecture of democracy*. Princeton Architectural Press. p. 23. ISBN 978-1-56898-134-5.

[45] maw/dpa (11 March 2008). "New Study Finds More Stasi Spooks" (http://www.spiegel.de/international/germany/0,1518,540771,00. html). *Spiegel Online - english site (www.spiegel.de/international)*. Der Spiegel. . Retrieved 30 October 2011. "189,000 people were informers for the Stasi - the former Communist secret police - when East Germany collapsed in 1989 - 15,000 more than previous studies had suggested. [...] about one in 20 members of the former East German Communist party, the SED, was a secret police informant."

[46] Colchester, Nico (1 January 2001). "D-mark day dawns" (http://www.ft.com/cms/s/2/ 504285c4-68b6-11da-bd30-0000779e2340,dwp_uuid=6f876a3c-e19f-11da-bf4c-0000779e2340.html). *Financial Times* (London). . Retrieved 19 March 2011.

[47] Protzman, Ferdinand (22 August 1989). "Westward Tide of East Germans Is a Popular No-Confidence Vote" (http://www.nytimes.com/ 1989/08/22/world/westward-tide-of-east-germans-is-a-popular-no-confidence-vote.html?pagewanted=all&src=pm). *www.nytimes.com*. The New York Times. . Retrieved 30 October 2011. "Behind the mass flight, Western experts say, is widespread and deepening disillusionment with the Honecker leadership's policies, particularly the refusal to consider the type of economic and political changes taking place elsewhere in Eastern Europe."

[48] "Gesetz zur Umsetzung des Beschlusses des Deutschen Bundestages vom 20. Juni 1991 zur Vollendung der Einheit Deutschlands" (http:// bundesrecht.juris.de/berlin_bonng/index.html) (in German). Bundesministerium der Justiz. 26 April 1994. . Retrieved 19 April 2011.

[49] "Brennpunkt: Hauptstadt-Umzug" (http://www.focus.de/panorama/boulevard/brennpunkt-hauptstadt-umzug_aid_175751.html) (in German). *Focus* (Munich). 12 April 1999. . Retrieved 19 March 2011.

[50] Dempsey, Judy (31 October 2006). "Germany is planning a Bosnia withdrawal" (http://www.nytimes.com/2006/10/31/world/europe/ 31iht-germany.3343963.html). *International Herald Tribune* (Paris). . Retrieved 7 May 2011.

[51] Merz, Sebastian (November 2007). "Still on the way to Afghanistan? Germany and its forces in the Hindu Kush" (http://www.sipri.org/ research/conflict/publications/merz) (PDF). Stockholm International Peace Research Institute. pp. 2, 3. . Retrieved 16 April 2011.

[52] "World Factbook" (https://www.cia.gov/library/publications/the-world-factbook/geos/gm.html). CIA. . Retrieved 26 March 2011.

[53] "Climate in Germany" (http://www.germanculture.com.ua/library/facts/bl_climate.htm). GermanCulture. . Retrieved 26 March 2011.

[54] "Terrestrial Ecoregions" (http://wwf.panda.org/about_our_earth/ecoregions/ecoregion_list/). WWF. . Retrieved 19 March 2011.

[55] Strohm, Kathrin (May 2010). "Arable farming in Germany" (http://www.agribenchmark.org/fileadmin/freefiles/ccc_2010/ Poster_Germany.pdf). Agri benchmark. . Retrieved 14 April 2011.

[56] Bekker, Henk (2005). *Adventure Guide Germany*. Hunter. p. 14. ISBN 9781588435033.

[57] "Zoo Facts" (http://www.americanzoos.info/Zoofacts.html). Zoos and Aquariums of America. . Retrieved 16 April 2011.

[58] "Der Zoologische Garten Berlin" (http://www.zoo-berlin.de/zoo/unternehmen/historie.html) (in German). Zoo Berlin. . Retrieved 19 March 2011.

[59] "Basic Law for the Federal Republic of Germany" (https://www.btg-bestellservice.de/pdf/80201000.pdf). *Deutscher Bundestag*. Btg-bestellservice. October 2010. . Retrieved 14 April 2011.

[60] "Christian Democratic Union/Christian Social Union" (http://countrystudies.us/germany/159.htm). U.S. Library of Congress. . Retrieved 26 March 2011.

[61] "Federal Constitutional Court" (http://www.bundesverfassungsgericht.de/en/index.html). Bundesverfassungsgericht. . Retrieved 26 March 2011.

[62] "Völkerstrafgesetz Teil 1 Allgemeine Regelungen" (http://bundesrecht.juris.de/vstgb/BJNR225410002. html#BJNR225410002BJNG000200000) (in German). Bundesministerium der Justiz. . Retrieved 19 April 2011.

[63] "§ 2 Strafvollzugsgesetz" (http://www.gesetze-im-internet.de/stvollzg/__2.html) (in German). Bundesministerium der Justiz. . Retrieved 26 March 2011.

[64] Jehle, Jörg-Martin; German Federal Ministry of Justice (2009). *Criminal Justice in Germany* (http://books.google.com/
      books?id=-V-ng-8jOoQC&pg=PA23). Forum-Verlag. p. 23. ISBN 9783936999518. .

[65] Casper, Gerhard; Zeisel, Hans (January 1972). "Lay Judges in the German Criminal Courts". *Journal of Legal Studies* **1** (1): 141.
      JSTOR 724014.

[66] The individual denomination is either *Land* [state], *Freistaat* [free state] or *Freie (und) Hansestadt* [free (and) Hanseatic city].
      "The Federal States" (http://www.bundesrat.de/nn_11006/EN/organisation-en/laender-en/laender-en-node.html?__nnn=true).
      www.bundesrat.de. Bundesrat of Germany. . Retrieved 17 July 2011.
      "Amtliche Bezeichnung der Bundesländer [Official denomination of federated states]" (http://www.auswaertiges-amt.de/DE/Infoservice/
      Terminologie/Bundeslaender/Uebersicht_node.html) (in German) (PDF; download file „Englisch"). *www.auswaertiges-amt.de*. Federal
      Foreign Office. . Retrieved 22 October 2011.

[67] "Example for state constitution: "Constitution of the Land of North Rhine-Westphalia"" (http://www.landtag.nrw.de/portal/WWW/
      GB_I/I.7/Europa/Wissenswertes/English_information/North_Rhine_Westphalia_Constitution_revised.jsp). Landtag (state assembly) of
      North Rhine-Westphalia. . Retrieved 17 July 2011.

[68] "Kreisfreie Städte und Landkreise nach Fläche und Bevölkerung 31.12.2009" (http://www.destatis.de/jetspeed/portal/cms/Sites/
      destatis/Internet/DE/Content/Statistiken/Regionales/Gemeindeverzeichnis/Administrativ/Aktuell/04__KreiseAktuell,property=file.xls)
      (in German) (XLS). Statistisches Bundesamt Deutschland. October 2010. . Retrieved 26 September 2011.

[69] "German Missions Abroad" (http://www.auswaertiges-amt.de/EN/AAmt/Auslandsvertretungen/Uebersicht_node.html). German
      Federal Foreign Office. . Retrieved 26 March 2011.

[70] "The EU budget 2011 in figures" (http://ec.europa.eu/budget/figures/2011/2011_en.cfm). European Commission. . Retrieved 6 May
      2011.

[71] "United Nations regular budget for the year 2011" (http://www.un.org/ga/search/view_doc.asp?symbol=ST/ADM/SER.B/824). UN
      Committee on Contributions. . Retrieved 6 May 2011.

[72] "Declaration by the Franco-German Defence and Security Council" (http://www.ambafrance-uk.org/
      Declaration-by-the-Franco-German,4519.html). French Embassy UK. 13 May 2004. . Retrieved 19 March 2011.

[73] Freed, John C. (4 April 2008). "The leader of Europe? Answers an ocean apart" (http://www.nytimes.com/2008/04/04/world/europe/
      04iht-poll.4.11666423.html). *The New York Times*. . Retrieved 28 March 2011.

[74] "Aims of German development policy" (http://www.bmz.de/en/index.html). Federal Ministry for Economic Cooperation and
      Development. 10 April 2008. . Retrieved 26 March 2011.

[75] "Net Official Development Assistance 2009" (http://www.oecd.org/dataoecd/17/9/44981892.pdf). OECD. . Retrieved 26 March 2011.

[76] "Speech by Chancellor Angela Merkel to the United Nations General Assembly" (http://www.bundesregierung.de/nn_6566/Content/
      EN/Reden/2010/2010-09-21-merkel-mdg-gipfel.html). *Die Bundesregierung*. 21 September 2010. . Retrieved 18 March 2011.

[77] Harrison, Hope (2004). "American détente and German ostpolitik, 1969–1972" (http://www.ghi-dc.org/files/publications/bu_supp/
      supp1/supp-01_005.pdf). *Bulletin Supplement* (German Historical Institute) **1**. . Retrieved 26 March 2011.

[78] "Germany's New Face Abroad" (http://www.dw-world.de/dw/article/0,2144,1741310,00.html). *Deutsche Welle*. 14 October 2005. .
      Retrieved 26 March 2011.

[79] "Ready for a Bush hug?" (http://www.economist.com/node/7141311?story_id=7141311). *The Economist*. 6 July 2006. . Retrieved 19
      March 2011.

[80] "U.S.-German Economic Relations Factsheet" (http://germany.usembassy.gov/germany/img/assets/9336/econ_factsheet_may2006.
      pdf). U.S. Embassy in Berlin. May 2006. . Retrieved 26 March 2011.

[81] "Background paper on SIPRI military expenditure data, 2010" (http://www.sipri.org/research/armaments/milex/factsheet2010).
      Stockholm International Peace Research Institute. 11 April 2011. . Retrieved 7 May 2011.

[82] "Grundgesetz für die Bundesrepublik Deutschland, Artikel 65a,87,115b" (http://www.gesetze-im-internet.de/bundesrecht/gg/gesamt.
      pdf) (in German). Bundesministerium der Justiz. . Retrieved 19 March 2011.

[83] "Die Stärke der Streitkräfte" (http://www.bundeswehr.de/portal/a/bwde/!ut/p/c4/
      DcmxDYAwDATAWVgg7unYAugc8kSWI4OMIesTXXm002D8SeWQy7jRStshc-4p94L0hENCnXEGUvXXSuMKG8FwBd26TD9uIZiT/)
      (in German). Bundeswehr. . Retrieved 5 June 2011.

[84] "Ausblick: Die Bundeswehr der Zukunft" (http://www.bundeswehr.de/portal/a/bwde/!ut/p/c4/
      04_SB8K8xLLM9MSSzPy8xBz9CP3I5EyrpHK9pPKUVL3ikqLUzJLsosTUtJJUvbzU0vTU4pLEnJLSvHRUuYKcxDygoH5BtqMiAMTJdF8!/
) (in German). Bundeswehr. . Retrieved 5 June 2011.

[85] "Einsatzzahlen – Die Stärke der deutschen Einsatzkontingente" (http://www.bundeswehr.de/portal/a/bwde/einsaetze/
      einsatzzahlen?yw_contentURL=/C1256EF4002AED30/W264VFT2439INFODE/content.jsp) (in German). Bundeswehr. . Retrieved 14
      April 2011.

[86] Connolly, Kate (22 November 2010). "Germany to abolish compulsory military service" (http://www.guardian.co.uk/world/2010/nov/
      22/germany-abolish-compulsory-military-service). *The Guardian* (UK). . Retrieved 7 April 2011.

[87] Pidd, Helen (16 March 2011). "Marching orders for conscription in Germany, but what will take its place?" (http://www.guardian.co.uk/
      world/2011/mar/16/conscription-germany-army). *The Guardian* (UK). . Retrieved 7 April 2011.

[88] "Frauen in der Bundeswehr" (http://www.bundeswehr.de/portal/a/bwde/!ut/p/c4/
      FcwxEoUgGgDAXAE0l6O0_x1YZ5QMSMEp2In-urs_3STC_FXzKqHIqdRpqi9KG50BK7qxpL3Qy8VHbZbk07MqtbDDerF_WJzYdGv286DbmAJj26iLgynaUN
) (in German). Bundeswehr. . Retrieved 14 April 2011.

[89] Norris, Floyd (20 February 2010). "A Shift in the Export Powerhouses" (http://www.nytimes.com/2010/02/20/business/economy/ 20charts.html). *The New York Times.* . Retrieved 27 March 2011.

[90] "CPI 2009 table" (http://www.transparency.org/policy_research/surveys_indices/cpi/2009/cpi_2009_table). Transparency International. . Retrieved 27 March 2011.

[91] "The Innovation Imperative in Manufacturing: How the United States Can Restore Its Edge" (http://www.bcg.com/documents/file15445. pdf). Boston Consulting Group. March 2009. . Retrieved 19 March 2011.

[92] "Gross domestic product (2009)" (http://siteresources.worldbank.org/DATASTATISTICS/Resources/GDP.pdf). *The World Bank: World Development Indicators database.* World Bank. 27 September 2010. . Retrieved 1 January 2011.
      Field listing – GDP (official exchange rate) (https://www.cia.gov/library/publications/the-world-factbook/fields/2195.html), CIA World Factbook

[93] "German Economy Experiences Record Growth in 2010" (http://www.germany.info/Vertretung/usa/en/__pr/P__Wash/2011/01/ 12__GDP__2010__PR.html) (Press release). German Missions in the United States. 12 January 2011. . Retrieved 27 March 2011.

[94] "Wind Power" (http://web.archive.org/web/20061210163253/http://www.german-renewable-energy.com/Renewables/Navigation/ Englisch/wind-power.html). Federal Ministry of Economics and Technology. Archived from the original (http://www. german-renewable-energy.com/Renewables/Navigation/Englisch/wind-power.html) on 2006-12-10. . Retrieved 27 March 2011.

[95] UFI, the Global Association of the Exhibition Industry (2008). "Euro Fair Statistics 2008" (http://www.auma.de/_pages/d/ 16_Download/download/FKM/EuroFairStatistics_2008.pdf). AUMA Ausstellungs- und Messe-Ausschuss der Deutschen Wirtschaft e.V.. p. 12. . Retrieved 24 September 2011.

[96] Andrews, Edmund L. (1 January 2002). "Germans Say Goodbye to the Mark, a Symbol of Strength and Unity" (http://www.nytimes.com/ 2002/01/01/world/germans-say-goodbye-to-the-mark-a-symbol-of-strength-and-unity.html). *The New York Times.* . Retrieved 18 March 2011.

[97] Taylor Martin, Susan (28 December 1998). "On Jan. 1, out of many arises one Euro". *St. Petersburg Times*: p. National, 1.A.

[98] Berg, S.; Winter, S.; Wassermann, A. (5 September 2005). "The Price of a Failed Reunification" (http://www.spiegel.de/international/ spiegel/0,1518,373639,00.html). *Spiegel Online.* . Retrieved 28 November 2006.

[99] Kulish, Nicholas (19 June 2009). "In East Germany, a Decline as Stark as a Wall" (http://www.nytimes.com/2009/06/19/world/europe/ 19germany.html). *The New York Times.* . Retrieved 27 March 2011.

[100] "Germany agrees on 50-billion-euro stimulus plan" (http://www.france24.com/en/ 20090106-germany-agrees-new-50-billion-euro-stimulus-plan). *France 24.* 6 January 2009. . Retrieved 27 March 2011.

[101] "The 100 Top Brands 2010" (http://www.interbrand.com/en/best-global-brands/best-global-brands-2008/best-global-brands-2010. aspx). Interbrand. . Retrieved 27 March 2011.

[102] Gavin, Mike (23 September 2010). "Germany Has 1,000 Market-Leading Companies, Manager-Magazin Says" (http://www. businessweek.com/news/2010-09-23/germany-has-1-000-market-leading-companies-manager-magazin-says.html). *Businessweek* (New York). . Retrieved 27 March 2011.

[103] "Global 500: Countries – Germany" (http://money.cnn.com/magazines/fortune/global500/2010/countries/Germany.html). *Forbes.* 26 July 2010. . Retrieved 27 March 2011.

[104] "Autobahn-Temporegelung" (http://www.presse.adac.de/standpunkte/Verkehr/Autobahn_Temporegelung. asp?active1=tcm:11-18784-4) (in German) (Press release). ADAC. June 2010. . Retrieved 19 March 2011.

[105] "Geschäftsbericht 2006" (http://web.archive.org/web/20070809140315/http://www.db.de/site/bahn/de/unternehmen/ investor__relations/finanzberichte/geschaeftsbericht/geschaeftsbericht__2006.html) (in German). Deutsche Bahn. Archived from the original (http://www.db.de/site/bahn/de/unternehmen/investor__relations/finanzberichte/geschaeftsbericht/geschaeftsbericht__2006. html) on 2007-08-09. . Retrieved 27 March 2011.

[106] "Airports in Germany" (http://www.aircraft-charter-world.com/airports/europe/germany.htm). Air Broker Center International. . Retrieved 16 April 2011.

[107] "Overview/Data: Germany" (http://www.eia.gov/countries/country-data.cfm?fips=GM). U.S. Energy Information Administration. 30 June 2010. . Retrieved 19 April 2011.

[108] "Energy imports, net (% of energy use)" (http://data.worldbank.org/indicator/EG.IMP.CONS.ZS). The World Bank Group. . Retrieved 18 April 2011.

[109] Reuters Berlin (7 July 2010). "* Environment * Renewable energy Germany targets switch to 100% renewables for its electricity by 2050" (http://www.guardian.co.uk/environment/2010/jul/07/germany-renewable-energy-electricity). *The Guardian* (UK). . Retrieved 18 April 2011.

[110] "Primärenergieverbrauch nach Energieträgern" (http://www.bmwi.de/BMWi/Redaktion/Binaer/Energiedaten/ energiegewinnung-und-energieverbrauch2-primaerenergieverbrauch.xls) (in German). Bundesministerium für Wirtschaft und Technologie. December 2010. . Retrieved 18 April 2011.

[111] "Germany split over green energy" (http://news.bbc.co.uk/2/hi/europe/4295389.stm). *BBC News.* 25 February 2005. . Retrieved 27 March 2011.

[112] "Deutschland erfüllte 2008 seine Klimaschutzverpflichtung nach dem Kyoto-Protokoll" (http://www.kst.portalu.de/dokumente/ Aktuelles/Inventar-Bericht_Treibhausgas_2008.pdf) (in German) (Press release). Umweltbundesamt. 1 February 2010. . Retrieved 19 March 2011.

[113] "Germany greenest country in the world" (http://articles.timesofindia.indiatimes.com/2008-06-21/pollution/
       27752791_1_energy-security-global-energy-germany). *The Times of India* (New Delhi). 21 June 2008. . Retrieved 26 March 2011.

[114] "Treibhausgas-Emissionen der Europäischen Union ..." (http://www.umweltbundesamt-daten-zur-umwelt.de/umweltdaten/public/
       document/downloadImage.do?ident=18198) (in German) (PDF). Umweltbundesamt. 2009. . Retrieved 10 May 2011.

[115] "Germany's Technological Performance" (http://www.bmbf.de/en/1869.php). Federal Ministry of Education and Research. 11 January
       2007. . Retrieved 21 August 2011.

[116] "Nobel Prize" (http://nobelprize.org/). Nobelprize.org. . Retrieved 27 March 2011.

[117] "Swedish academy awards" (http://www.sciencenews.org/view/generic/id/63944/title/Swedish_academy_awards). *ScienceNews*. .
       Retrieved 1 October 2010.

[118] National Science Nobel Prize shares 1901–2009 by citizenship at the time of the award (http://www.idsia.ch/~juergen/sci.html) and by
       country of birth (http://www.idsia.ch/~juergen/scinat.html). From Schmidhuber, J. (2010). "Evolution of National Nobel Prize Shares in
       the 20th century" (http://www.idsia.ch/~juergen/nobelshare.html). . Retrieved 27 March 2011.

[119] Roberts, J. M. (2002). *The New Penguin History of the World*. Allen Lane. p. 1014. ISBN 9780713996111.

[120] "The First Nobel Prize" (http://www.dw-world.de/dw/article/0,,5984670,00.html). *Deutsche Welle*. 8 September 2010. . Retrieved 27
       March 2011.

[121] "Gottfried Wilhelm Leibniz Prize" (http://web.archive.org/web/20080621091621/http://www.dfg.de/en/research_funding/
       scientific_prizes/gw_leibniz_prize.html). DFG. Archived from the original (http://www.dfg.de/en/research_funding/scientific_prizes/
       gw_leibniz_prize.html) on 2008-06-21. . Retrieved 27 March 2011.

[122] Bianchi, Luigi. "The Great Electromechanical Computers" (http://www.yorku.ca/lbianchi/sts3700b/lecture17a.html). York University.
       . Retrieved 17 April 2011.

[123] "The Zeppelin" (http://www.centennialofflight.gov/essay/Lighter_than_air/zeppelin/LTA8.htm). U.S. Centennial of Flight
       Commission. . Retrieved 27 March 2011.

[124] "Historical figures in telecommunications" (http://www.itu.int/aboutitu/HistoricalFigures.html). International Telecommunication
       Union. 14 January 2004. . Retrieved 27 March 2011.

[125] Roland Berger Strategy Consultants: *Green Growth, Green Profit – How Green Transformation Boosts Business* Palgrave Macmillan, New
       York 2010, ISBN 978-0-230-28543-9

[126] "Country Comparison :: Population" (https://www.cia.gov/library/publications/the-world-factbook/rankorder/2119rank.
       html?countryName=Lebanon&countryCode=le®ionCode=me&rank=126). CIA. . Retrieved 26 June 2011.

[127] Destatis. "Durchschnittliche Kinderzahl 2008 in den neuen Ländern angestiegen" (http://www.destatis.de/jetspeed/portal/cms/Sites/
       destatis/Internet/DE/Presse/pm/2010/01/PD10__034__12641,templateId=renderPrint.psml) (in German). . Retrieved 28 March 2011.

[128] "Demographic Transition Model" (http://geographyfieldwork.com/DemographicTransition.htm). Barcelona Field Studies Centre. 27
       September 2009. . Retrieved 28 March 2011.

[129] "Im Jahr 2060 wird jeder Siebente 80 Jahre oder älter sein" (http://www.destatis.de/jetspeed/portal/cms/Sites/destatis/Internet/DE/
       Presse/pm/2009/11/PD09__435__12411,templateId=renderPrint.psml) (in German) (Press release). Destatis. 18 November 2009. .
       Retrieved 12 February 2011.
       Details on the methodology, detailed tables, etc. are provided at "Bevölkerungsentwicklung in Deutschland bis 2060" (http://www.destatis.
       de/jetspeed/portal/cms/Sites/destatis/Internet/DE/Presse/abisz/Bevoelkerungsvorausberechnung,templateId=renderPrint.psml) (in
       German). Statistisches Bundesamt. 18 November 2009. . Retrieved 15 February 2011.

[130] (in German) (PDF) *Bevölkerung und Erwerbstätigkeit:Bevölkerung mit Migrationshintergrund – Ergebnisse des Mikrozensus 2009
       [Population and employment: Population with migrant background – Results of the 2009 microcensus]* (http://www.destatis.de/jetspeed/
       portal/cms/Sites/destatis/Internet/DE/Content/Publikationen/Fachveroeffentlichungen/Bevoelkerung/MigrationIntegration/
       Migrationshintergrund2010220097004,property=file.pdf). Fachserie 1 Reihe 2.2. Statistisches Bundesamt. 14 July 2010. pp. 6–8.
       Artikelnummer: 2010220097004. . Retrieved 7 May 2011.
       The Federal Statistical Office defines persons with a migrant background as all persons who migrated to the present area of the Federal
       Republic of Germany after 1949, plus all foreign nationals born in Germany and all persons born in Germany as German nationals with at
       least one parent who migrated to Germany or was born in Germany as a foreign national.

[131] "International Migration 2006" (http://www.un.org/esa/population/publications/2006Migration_Chart/Migration2006.pdf). UN
       Department of Economic and Social Affairs. . Retrieved 18 March 2011.

[132] "Germany" (http://www.focus-migration.de/Germany.1509.0.html?&L=1). Focus-Migration. . Retrieved 28 March 2011.

[133] "20% of Germans have immigrant roots". *Burlington Free Press*: p. 4A. 15 July 2010.

[134] "Bevölkerung nach Migrationshintergrund" (http://www.destatis.de/jetspeed/portal/cms/Sites/destatis/Internet/DE/Content/
       Statistiken/Bevoelkerung/MigrationIntegration/Migrationshintergrund/Tabellen/Content100/
       MigrationshintergrundStaatsangehoerigkeit,templateId=renderPrint.psml) (in German). German Federal Statistical Office. . Retrieved 28
       March 2011.

[135] "Fewer Ethnic Germans Immigrating to Ancestral Homeland" (http://www.migrationinformation.org/Feature/display.cfm?ID=201).
       Migration Information Source. February 2004. . Retrieved 28 March 2011.

[136] "Regionales Monitoring 2008 – Daten und Karten zu den Europäischen Metropolregionen (EMR) in Deutschland" (http://www.
       deutsche-metropolregionen.org/fileadmin/ikm/IKM-Veroeffentlichungen/IKM-Monitoring2008_lite.pdf) (in German). Bundesamt für
       Bauwesen und Raumordnung. 2008. p. 10. . Retrieved 11 June 2011.

[137] "EKD-Statistik: Christen in Deutschland 2007" (http://www.ekd.de/statistik/mitglieder.html) (in German). Evangelische Kirche in Deutschland. . Retrieved 28 March 2011.

[138] "Konfessionen in Deutschland" (http://fowid.de/fileadmin/datenarchiv/Religionszugehoerigkeit_Bevoelkerung__1950-2008.pdf) (in German). Fowid. 9 September 2009. . Retrieved 28 March 2011.

[139] "Chapter 2: Wie viele Muslime leben in Deutschland?" (http://www.bmi.bund.de/cae/servlet/contentblob/566008/publicationFile/31710/vollversion_studie_muslim_leben_deutschland_.pdf;jsessionid=6B8CD26E2AC179111AF4F75650B84B1A) (in German) (PDF). *Muslimisches Leben in Deutschland*. Bundesamt für Migration und Flüchtlinge. June 2009. pp. 80, 97. ISBN 978-3-9812115-1-1. . Retrieved 28 March 2011.

[140] "Religionen in Deutschland: Mitgliederzahlen" (http://www.remid.de/remid_info_zahlen.htm) (in German). Religionswissenschaftlicher Medien- und Informationsdienst. 31 October 2009. . Retrieved 28 March 2011.

[141] Blake, Mariah (10 November 2006). "In Nazi cradle, Germany marks Jewish renaissance" (http://www.csmonitor.com/2006/1110/p25s02-woeu.html). *Christian Science Monitor*. . Retrieved 28 March 2011.

[142] Schnabel, U. (15 March 2007). "Buddhismus Eine Religion ohne Gott" (http://www.zeit.de/2007/12/Buddhismus) (in German). *Die Zeit* (Hamburg). . Retrieved 19 March 2011.

[143] European Commission (2006). "Special Eurobarometer 243: Europeans and their Languages (Survey)" (http://ec.europa.eu/public_opinion/archives/ebs/ebs_243_en.pdf). Europa (web portal). . Retrieved 28 March 2011.
       European Commission (2006). "Special Eurobarometer 243: Europeans and their Languages (Executive Summary)" (http://ec.europa.eu/public_opinion/archives/ebs/ebs_243_sum_en.pdf). Europa (web portal). . Retrieved 28 March 2011.

[144] European Commission (2004). "Many tongues, one family. Languages in the European Union" (http://ec.europa.eu/publications/booklets/move/45/en.pdf). Europa (web portal). . Retrieved 28 March 2011.

[145] "Sprechen Sie Deutsch?" (http://www.economist.com/node/15731354). *The Economist*. 18 March 2010. . Retrieved 16 April 2011.

[146] Grotlüschen, Anke; Riekmann, Wibke (2011). "leo.- Level One Survey. Presseheft" (http://www.vhs-sachsen.de/uploads/media/Executive_Summary_leo_v8_Journalismusfassung.pdf) (in German). University of Hamburg. . Retrieved 15 April 2011.

[147] "Country profile: Germany" (http://lcweb2.loc.gov/frd/cs/profiles/Germany.pdf). Library of Congress. April 2008. . Retrieved 28 March 2011.

[148] "The Educational System in Germany" (http://academic.cuesta.edu/intlang/german/education.html). Cuesta College. 31 August 2002. . Retrieved 16 May 2011.

[149] "Top 100 World Universities" (http://web.archive.org/web/20080822124509/http://www.arwu.org/rank2008/ARWU2008_A(EN).htm). Academic Ranking of World Universities. Archived from the original (http://www.arwu.org/rank2008/ARWU2008_A(EN).htm) on 2008-08-22. . Retrieved 28 March 2011.

[150] "Tuition Fees at university in Germany" (http://www.studyineurope.eu/study-in-germany/admission/tuition-fees). StudyinEurope.eu. 2009. . Retrieved 19 March 2011.

[151] *Health Care Systems in Transition: Germany* (http://www.euro.who.int/__data/assets/pdf_file/0010/80776/E68952.pdf). European Observatory on Health Care Systems. 2000. p. 8. AMS 5012667 (DEU). . Retrieved 15 April 2011.

[152] "Core Health Indicators" (http://apps.who.int/whosis/database/core/core_select.cfm). World Health Organization. . Retrieved 6 June 2011.

[153] "Statistisches Bundesamt Deutschland – Herz-/Kreislauferkrankungen nach wie vor häufigste Todesursache" (http://www.destatis.de/jetspeed/portal/cms/Sites/destatis/Internet/DE/Presse/pm/2010/10/PD10__371__232,templateId=renderPrint.psml) (in German). Destatis.de. . Retrieved 2011-06-07.

[154] "Country Profile Germany" (http://lcweb2.loc.gov/frd/cs/profiles/Germany.pdf) (PDF). Library of Congress Federal Research Division. April 2008. . Retrieved 2011-05-07.
       This article may incorporate text from this source, which is in the public domain.

[155] "Topping the EU Fat Stats, Germany Plans Anti-Obesity Drive" (http://www.dw-world.de/dw/article/0,,2449356,00.html). Deutsche Welle. 20 April 2007. . Retrieved 28 March 2011.

[156] "Germany launches obesity campaign" (http://news.bbc.co.uk/2/hi/health/6639227.stm). BBC. 9 May 2007. . Retrieved 28 March 2011.

[157] Wasser, Jeremy (6 April 2006). "Spätzle Westerns" (http://www.spiegel.de/international/0,1518,410135,00.html). *Spiegel Online International*. . Retrieved 28 March 2011.

[158] "Unbelievable Multitude" (http://www.dw-world.de/dw/0,,8009,00.html). *Deutsche Welle*. . Retrieved 28 March 2011.

[159] "World Heritage Sites in Germany" (http://www.worldheritagesite.org/countries/germany.html). UNESCO. . Retrieved 3 October 2010.

[160] "Human Development Report 2010 Table 4 Gender Inequality Index" (http://hdr.undp.org/en/media/HDR_2010_EN_Tables_reprint.pdf). United Nations Development Programme. pp. 156–160. . Retrieved 20 April 2011.

[161] "Germany extends gay rights" (http://www.news24.com/World/News/Germany-extends-gay-rights-20041029). News24. 29 October 2004. . Retrieved 19 March 2011.

[162] Heckmann, Friedrich (2003). *The Integration of Immigrants in European Societies: national differences and trends of convergence*. Lucius & Lucius. pp. 51 ff. ISBN 978-3-8282-0181-1.

[163] "2010 Anholt-GfK Roper Nation Brands Index" (http://www.gfk.com/group/press_information/press_releases/006688/index.en.html) (Press release). GfK. 12 October 2010. . Retrieved 28 March 2011.

[164] "Views of US Continue to Improve in 2011 BBC Country Rating Poll" (http://www.worldpublicopinion.org/pipa/articles/views_on_countriesregions_bt/680.php?nid=&id=&pnt=680&lb=#ger). *Worldpublicopinion.org*. 7 March 2011. . Retrieved 28 March 2011.

[165] *A Dictionary of Architecture and Landscape Architecture*. Oxford University Press. 2006. p. 880. ISBN 0-19-860678-8.

[166] "Germany's flailing music industry seeks new talent" (http://www.dw-world.de/dw/article/0,,5200389,00.html). *Deutsche Welle*. 2 February 2010. . Retrieved 28 March 2011.

[167] Espmark, Kjell (3 December 1999). "The Nobel Prize in Literature" (http://nobelprize.org/nobel_prizes/literature/articles/espmark/index.html). Nobelprize.org. . Retrieved 28 March 2011.

[168] "Land of ideas" (http://land-der-ideen.matrix.de/CDA/facts_printing,4563,0,,en.html). Land-der-ideen.matrix.de. . Retrieved 19 March 2011.

[169] Weidhaas, Peter; Gossage, Carolyn; Wright, Wendy A. (2007). *A History of the Frankfurt Book Fair*. Dundurn Press Ltd.. pp. 11 ff. ISBN 9781550027440.

[170] Searle, John (1987). "Introduction". *The Blackwell Companion to Philosophy*. Wiley-Blackwell.

[171] Bordwell, David; Thompson, Kristin (2003) [1994]. "The Introduction of Sound". *Film History: An Introduction* (2nd ed.). McGraw-Hill. p. 204. ISBN 978-0-07-115141-2.

[172] "Rainer Werner Fassbinder" (http://www.fassbinderfoundation.de/node.php/en/home). Fassbinder Foundation. . Retrieved 28 March 2011.

[173] "2006 FIAPF accredited Festivals Directory" (http://www.fiapf.org/pdf/2006accreditedFestivalsDirectory.pdf). International Federation of Film Producers Associations. . Retrieved 28 March 2011.

[174] "Awards:Das Leben der Anderen" (http://www.imdb.com/title/tt0405094/awards). IMDb. . Retrieved 28 March 2011.

[175] "Country profile: Germany" (http://news.bbc.co.uk/2/hi/europe/country_profiles/1047864.stm). *BBC News*. . Retrieved 28 March 2011.

[176] "Guide to German Sausages & Meat Products" (http://www.foodfromgermany.org/consumer/facts/guidetosausages.cfm). German Foods North America. . Retrieved 11 May 2011.

[177] "Germany Country Profiles for Organic Agriculture" (http://www.fao.org/organicag/display/work/display_2.asp?country=DEU&lang=en&disp=summaries). Food and Agriculture Organization. . Retrieved 6 May 2011.

[178] "Europe's largest beer market 2006" (http://www.royalunibrew.com/Default.aspx?ID=266). Royal Unibrew. . Retrieved 28 March 2011.

[179] "Schnitzel Outcooks Spaghetti in Michelin Guide" (http://www.dw-world.de/dw/article/0,2144,2914502,00.html). *Deutsche Welle* (Bonn). 15 November 2007. . Retrieved 28 March 2011.

[180] "German cuisine beats Italy, Spain in gourmet stars" (http://in.reuters.com/article/2007/11/14/us-germany-food-idINL1447732320071114). *Reuters*. 28 March 2011. . Retrieved 19 March 2011.

[181] "Germany Info: Culture & Life: Sports" (http://www.germany.info/relaunch/culture/life/sports.html). Germany Embassy in Washington, D.C. . Retrieved 28 March 2011.

[182] Ornstein, David (23 October 2006). "What we will miss about Michael Schumacher" (http://www.guardian.co.uk/sport/2006/oct/23/formulaone.sport). *The Guardian* (UK). . Retrieved 19 March 2011.

[183] "Beijing 2008 Medal Table" (http://www.olympic.org/medallists-results?athletename=&category=343488&games=1333952&sport=&event=&mengender=false&womengender=false&mixedgender=false&teamclassification=false&individualclassification=false&continent=&country=&goldmedal=true&silvermedal=true&bronzemedal=true&worldrecord=false&olympicrecord=false&targetresults=true). International Olympic Committee. . Retrieved 19 March 2011.

[184] "Turin 2006 Medal Table" (http://www.olympic.org/medallists-results?athletename=&category=343486&games=1334152&sport=&event=&mengender=false&womengender=false&mixedgender=false&teamclassification=false&individualclassification=false&continent=&country=&goldmedal=true&silvermedal=true&bronzemedal=true&worldrecord=false&olympicrecord=false&targetresults=true&sortorder=medal&sortorder=country). International Olympic Committee. . Retrieved 19 March 2011.

Work cited

Fulbrook, Mary (1991). *A Concise History of Germany*. Cambridge University Press. ISBN 9780521368360.

External links

- deutschland.de (http://www.deutschland.de/en/home-page.html?tx_fdfxyaml_pi1) – Official Germany portal (non-profit)
- Official site of German Chancellor (http://www.bundeskanzlerin.de/Webs/BK/En/Homepage/home.html)
- Deutsche Welle (http://www.dw-world.de/) – Germany's international broadcaster
- Germany (http://ucblibraries.colorado.edu/govpubs/for/germany.htm) at *UCB Libraries GovPubs*
- Germany (http://www.dmoz.org/Regional/Europe/Germany/) at the Open Directory Project

Wikimedia Atlas of Germany

- Facts about Germany (http://www.tatsachen-ueber-deutschland.de/en/) – by the German Federal Foreign Office
- Destatis.de (http://www.destatis.de/jetspeed/portal/cms/Sites/destatis/Internet/EN/Navigation/Homepage__NT.psml) – Federal Statistical Office Germany
- OpenStreetMap has geographic data related to Germany (http://www.openstreetmap.org/browse/relation/51477)

Article Sources and Contributors

Waltenhausen *Source*: http://en.wikipedia.org/w/index.php?title=Waltenhausen *Contributors*: Agathoclea, DerBorg, Ksnow, Leit, Markussep, Mimihitam, 2 anonymous edits

Municipalities of Germany *Source*: http://en.wikipedia.org/w/index.php?title=Municipalities_of_Germany *Contributors*: Acntx, Agathoclea, Carlossuarez46, Chanheigeorge, David Kernow, DerBorg, Ehrlich91, Mschiffler, Poulpy, Reinyday, Roke, Sardanaphalus, Stemonitis, Sundar1, Tobias Conradi, Valentinian, Worobiew, Yngvadottir, 3 anonymous edits

Regierungsbezirk *Source*: http://en.wikipedia.org/w/index.php?title=Regierungsbezirk *Contributors*: (, Agathoclea, Ahoerstemeier, Andrew c, Angr, Badanedwa, Bermicourt, CalJW, Capitonym, Chanheigeorge, Crusoe8181, David Parker, Delirium, DerBorg, Djmutex, Docu, Donald Albury, Dvavasour, Ehrlich91, Elapsed, Eugene van der Pijll, Exec, Frokor, Funandtrvl, Gereonmc, GregorB, Ingolfson, JPosten, JeLuF, KaiKemmann, Kelisi, Kwamikagami, Lfh, LilHelpa, Ludger1961, Marek69, MegA, Meursault2004, Mrmuk, Numbo3, Olessi, Picapica, Piotrus, Ratzer, Reywas92, Rje, Robofish, Rossami, Royboycrashfan, Saintswithin, Sandman, Timwi, Tom, Ulf Heinsohn, Unyoyega, Worobiew, Δ, 30 anonymous edits

Bad Wiessee *Source*: http://en.wikipedia.org/w/index.php?title=Bad_Wiessee *Contributors*: 52 Pickup, Aadelse, Attilios, Bigweeboy, Ksnow, Louis3ham, Markussep, MayaSimFan, Numbo3, OldakQuill, Pjamescowie, Rich Farmbrough, SimonP, SusanLarson, Woohookitty, 6 anonymous edits

Miesbach (district) *Source*: http://en.wikipedia.org/w/index.php?title=Miesbach_%28district%29 *Contributors*: Ahoerstemeier, Angusmclellan, Baldhur, Bermicourt, CharlotteWebb, GDuwen, Good Olfactory, Green caterpillar, Hokanomono, Imars, Markus.zhang, Markussep, Mishuletz, Net-net, Numbo3, Rich Farmbrough, Template namespace initialisation script, 8 anonymous edits

Bavaria *Source*: http://en.wikipedia.org/w/index.php?title=Bavaria *Contributors*: (, 0, 10 year old boy, 1000Faces, 111Alleskoenner, 128.196.133.xxx, 159753, 16@r, 2help, 52 Pickup, 842U, 87v7t76fc4iguwevf7657436253yd4fug754ws67dtfugiy67t8576, @xel, A history of the modern world, AAGUNMAN, ACV777, Abisch, Achates, Adam78, Adolphus79, Adrian J. Hunter, Agathoclea, Agne27, Ahoerstemeier, Ahuffski10, Akkolon, Alarob, Aleenf1, Alekjds, Alensha, Alexander Domanda, AlexiusHoratius, AllPeopleUnite, Alvaro, Amatulic, Andi1995, Andrew0921, AndrewRT, Andykentisi, Anger22, Angr, Animum, AnnaFrance, Anrod, Antandrus, Ardel Murzighiz, ArtVandelay13, Arthur Holland, Ask123, Attilios, AuburnPilot, Auslli, Avalon, AxSkov, BMWLover80, BYF, Baldhur, Baridiah, Bavarian Hiasl, Bender235, Betterart, BigMike718, Bigbossfarin, BilCat, Billinghurst, Bit swizzler, Blablablob, Bloodofox, Bloomfield, Bobblewik, Bobo192, Bogey97, Bomac, Boson, Brabban, Brion VIBBER, Buaidh, Burzmali, C.Kent87, Cacetudo, CalicoCatLover, Caltas, CambridgeBayWeather, Canyouhearmenow, Caponer, Cdecoro, Cfaerber, Chanheigeorge, Charles01, CharlotteWebb, Chd, Chekaz, Choster, CieloEstrellado, Cinnamon42, ClickRick, Cloachland, Closedmouth, Coemgenus, Colonel Cow, CommonsDelinker, Conscious, Conversion script, CopperSquare, Cosal, Coughinink, Craigzomack, DTOx, DWC LR, Dagbert, Dale Arnett, Dandelions, DanielCD, Darren Jowalsen, Darth Cookie Monster, Dbachmann, Dbcook256, Dcandeto, Dcb1995, Delta G, Demcs, Den fjättrade ankan, Denelson83, Denihilonihil, Der Eberswalder, Derekbd, Devoblue, Dewritech, Djchris1991, Djmutex, DocendoDiscimus, Doco, Docu, Domino theory, Donama, Drindl, Drunken Pirate, Dysepsion, ESkog, East718, Easter Monkey, Edcolins, Edward NZ, Einemnet, Ejosse1, Ekwos, El C, Elassint, Elockid, Elox, Energyfreezer, Epbr123, Erianna, Erichwiki, Erikeltic, Euchiasmus, Evadb, EverettColdwell, Fabartus, Fabianklenk, Fallschirmjäger, Fantasy, FastLizard4, Favonian, Felix.R.A., Ferkelparade, Fingers-of-Pyrex, Finnrind, Fishwristwatch, Fluffernutter, Fogeltje, FordPrefect42, Fractalizator, Frappyjohn, Fuddle, Fulcher, Future Perfect at Sunrise, Fxhomie, Fyyer, GCarty, GJR, GLGermann, Gadget850, Gaius Cornelius, Gamma, Gargolla, Garybel, Garzo, Gavia immer, Gecko78, General Jazza, Gentgeen, Ghosts&empties, Gilgamesh, Gilgamesh he, Gilliam, Glane23, Gnu1742, Gogafax, Gogo Dodo, Goldfishbutt, Gppande, Gr1st, Graham87, Grandiose, Grenavitar, Grimhelm, Grstain, Gryffindor, Gurch, Gwernol, H005, Hackercraft, Hadal, Hairy Dude, Hajenso, Halsteadk, Hanberke, HansWorst, Hardman1, Harrystein, Hashar, Hayden120, HeartofaDog, Henrygb, Hibernian, Highpriority, Hingster, Hinicome, Hixteilchen, Hmains, Hokanomono, Hoo man, Hristarhos, Hu12, Huntthetroll, Husnock, Hvn0413, Hydrargyrum, IEdML, IZAK, Iamiyouareyou, Ian Pitchford, Immanuel Giel, Imperator24, Int21h, Isis, Ithurts2poo, Ivirivi00, Ixfd64, J. 'mach' wust, J. Patrick Fischer, J.delanoy, JM.Beaubourg, JaGa, Jack.Danish, Jamesfett, Jamie Mercer, Jan eissfeldt, Jared Preston, Jasonuhl, Jcouser, Jeannedeba, Jebus989, Jeffman52001, Jeronimo, Jesster79, Jguad1, Jiang, Jj137, Jnestorius, Joewein, John K, John Vandenberg, JohnBlackburne, Jonas Liljeström, Jorunn, Joseph Solis in Australia, Joshuashearn, Jpgordon, Jpxt2000, Ju98 5, Juergen1, Justinc, Juxtaposit, Kait, Kate Phaye, Kato2k6, Kelisi, Kesac, Kgrad, Kingjeff, Kinneyboy90, Kirandinakaran, Kjamesh89, Kortnay08, KrakatoaKatie, Kristinae, Kritikos99, Krun, Ksademap, Kuru, Kusma, Kwamikagami, KyraVixen, LOL, Laca, Lacrimosus, Lars T., Lear 21, Leszek Jańczuk, Lethe, Lfh, Lightmouse, Lipedia, Lir, LorenzoB, Lswartz, Luatha, Lumpydog32, Lysy, MER-C, MONGO, MaGioZal, Mac c, MacTire02, Mackeriv, MadGeographer, Magnus Manske, Malleus Fatuorum, Mandarax, Manecke, Maproom, Maralex, Marcika, Marek69, Markofthebeast, Markus Kuhn, Markus Stamm, Markussep, Martarius, Martin451, Matthias Klostermayr, Mattissa, Maximus Rex, Mcferran, Meaghan, Melancholia i, Merovingian, Mevsfotw, Michael Hardy, MichaelTinkler, Michel rockstar, Mirage, Mirekmarut, Misza13, Mjquinn id, Monkeyblue, Moocha, Moombasa, Moomoocowsareawesome, Moonriddengirl, Morevisit, Morwen, Mr.K., MrBill47, Msedwick, Mxn, MyPasswordIs-FreeState, Nach0king, Nauticashades, NawlinWiki, Neolandes, Net-net, Nick Number, Nickshanks, Nico, Nigel Campbell, NoPuzzleStranger, Novil Ariandis, Number1schumacherfan, Numbo3, Ocee, Olessi, Olivier, Omicronpersei8, Omnieiunium, Orangemarlin, Oreo Priest, OwenBlacker, Paine Ellsworth, Pascal.Tesson, Patrick, Paulc206, Peeperman, Penwhale, Pepsitest, Peregrine981, Peter Isotalo, Peteweez, Phearson, Philip Trueman, Philopedia, Picapica, Piccolo Modificatore Laborioso, Pigpen, Pilotguy, Polozooza, Poosaviour93, Postcard Cathy, Prjaeger, Proofreader77, Prosbus, Pzich, Quadell, Qualiabavariae, Qxz, R'n'B, R-41, RRKennison, Rabauz, Rabbitrunning2, Rahuloof, Ramckay, Rarelibra, Rast5, Ratzer, Ravidreams, RayquazaDialgaWeird2210, Rbb l181, Rchamberlain, Rdb, Redsox00002, Reindra, RexNL, Reymma, Reywas92, Rhombus, Rich Farmbrough, Richie, Ringbang, Rivertorch, Rjbkearney, RonPab, Ronaldtheroofer, Rosenknospe, Rosh2610, Rrius, Ruhrjung, Rupertslander, Ruud64, RyanGerbil10, Rōnin, S3000, STB110, Saluton, Sam Blacketer, Sanchristobal76, Sandman, Santryl, Sarah, Sardanaphalus, Scarian, Schneelocke, SchuminWeb, SchweizerJunge, Scientizzle, Scriberius, Sd31415, Seatdistrict, SebastianHelm, Sejdb, Shanes, Shilkanni, SidP, Signalhead, SilverWiki, Silverhelm, Skibbledee, Skysmith, SlaskiAnanasek91, Sleigh, SluggoOne, Slysplace, Sm8900, SmartGuy, Smith2006, SnapSnap, Snowdog, Snoyes, SoWhy, Sobebunny, Son sonson, SpartanPhalanx8588, SpookyMulder, Srikeit, Srose, Steel, Stemonitis, Str1977, Sunderland06, Supertouch, Sverdrup, Symposiarch, Synchronism, TKD, Tabletop, Tagishsimon, Techman224, Tedius Zanarukando, Template namespace initialisation script, Tesi1700, Testsieger73, Texfive, The Behnam, The Photon, The Rambling Man, The Transhumanist, TheKMan, Thecheesykid, Thedjatclubrock, Themfromspace, Theo van der Zalm, Thunderboltz, Thylacoleo, TikiWiki, Timwi, Titoxd, Tobby72, Tommycw1, Tony1, Tpbradbury, Traceylily, Troglo, Ttownfeen, Tulkolahten, Twhite 88, Twp, Ulf Heinsohn, Ulritz, Unyoyega, Urban XII, Ussschrotti, UtherSRG, Uthlande, Valentinian, Valrith, Vanished User 1004, Vaquero100, Vdiest, Vdjj1960, Verazzano, Viewfinder, VodkaMartini, Walor, Whugel, Wi-king, Wilfried Derksen, Wilhelm meis, Winchelsea, Woohookitty, WordyGirl90, Writtenright, Wzwz, Yooden, Yossiea, Zebra848, Zeno Gantner, Ziko, ZingbatStew, Zinnmann, Zundark, Zzzzzozzzzz, 1021 anonymous edits

Germany *Source*: http://en.wikipedia.org/w/index.php?title=Germany *Contributors*: -- April, -Majestic-, 000peter, 02pollajo, 03md, 06alifar, 119, 123kelsey123, 1297, 14Adrian, 159753, 16@r, 1Winston, 247balla34, 334a, 52 Pickup, 7huizen, 8ung3st, A Werewolf, A bit iffy, A little insignificant, A-giau, ABCD, ABlockedUser, AJackl, ALE!, ANNRC, Aaker, Aaron, Aaron Einstein, Aaron Schulz, Aaronlws, Aazn, Abe Lincoln, Abonazzi, Abrech, Academic Challenger, Academicigbo, Acalamari, Acanon, Achates, Adam Bishop, Adam Lutostański, Adashiel, Adcva, Adherent of the Enlightenment 10.0, AdmRose, Aeusoes1, Agathoclea, Ahoerstemeier, Ahrarara, Ahuskay, Aitias, Aivazovsky, AjaxSmack, Ajenterp, Akanemoto, Akhristov, Akkolon, Akradecki, Alai, Alan De Smet, Alanmak, Alarics, Alberto3166, Alex S, Alex Usoltsev, Alex1011, Alexandru Busa, Alexandru Stanoi, Alfirin, Alfonsomedina1, Algebraist, Algont, Alison, Alixus, All N Ever, Allgermein, Allmightyduck, AllyUnion, AlmightyClam, Alopex, Alphachimp, Alphasinus, Alphax, Altenmann, Alterrabe, Amakuru, Amber-W-12, Amcaja, AmiDaniel, Amoreno65, Anaki72, Ancheta Wis, Andre Engels, Andrew Levine, Andrew0921, Andrewlp1991, Andries, Andrij Kursetsky, Andris, Andrwsc, Andy Marchbanks, Andyjsmith, Andypandy.UK, Anetode, Anger22, Angr, AniRaptor2001, Animum, Anjwalker, Ann O'nyme, Anna Lincoln, Anoldtreeok, Anonymous from the 21th century, Antandrus, Anteeru, Anthony3, Anthonymendoza, Antidote, Antientropic, Antonio Basto, Ap, Apeloverage, Apoc2400, Apokrif, Appleseed, Apuman, Aqwis, Aranherunar, Arathjp, Arbero, Arindam2011, Aris Katsaris, AndninoFromHell, Arnomane, Arnowaschk, Arsonal, Art LaPella, Artichoker, Arunnplus, Asams10, Asarelah, Asdfasdf1231234, Ashton1983, Askalan, Asnatu wiki, Asocall, AsusFan, Ataraxis1492, Athinaios, Aths, Atlant, Atlanticpuffin, Atlantik, Atoric, Atrix20, Attilios, Atwardow, AuburnPilot, Aude, Audiovideo, Aufbauten, Augustus Rookwood, Auntof6, Auric, Auriong, Autocracy, Avala, Avatarion, AvicAWB, Avicennasis, Avraham, Awolf002, AxG, Axegod12, Axel Phony, AxelBoldt, Axpde, Axt, Aye Carumba Fajita Pizza, Azate, Aznpryde, B1Ilycl3m3nt, BD2412, BGManofID, BJess, BRG, Babakexorramdin, Babbage, Backspace, Baldhur, Ballsontheroad, BalowStar, Bambuway, Banana04131, BanyanTree, Bapon, Barberio, Bardhylius, Barek, Baristarim, Barliner, Baronnet, Barras, Barryob, Barutazaru, Baseballnut290, Bastin, Battem, Baudrillard, Bazonka, Bbenjoe, Bbruchs, Bcorr, Beagel, Beetstra, Bejinhan, Beland, BenMerill, Bender235, Benefitov, BenjaT, Benne, Benster1510, Bento00, Benwildeboer, Benznine, Berkut, BernardaAlba, Bernardissimo, BernhardtP, Bert-25, Beta m, Betacommand, Betterworld, Bevo74, Bg007, Bhuck, BigBen212, Biker Biker, BilCat, Bilbo571, Bippy1990, Bjarki S, Bkell, Blablaaa, Blacha, BlackGothFaerie, Blackjack48, Blacklite, Blake24, Blake3522, BlazeTheMovieFan, Bleh999, Bletch, Blinder Seher, Blizzardstep0, Bloodofox, Blue-Haired Lawyer, Blue520, BlueMars, Bluemask, Bluezy, Blur4760, Bml, Bob rulz, Bob5409, Bob889900, Bobblewik, Bobby122, Bobet, Bobo192, Bocafan76, Bohuiginn, Boothy443, BorgHunter, Bornsommer, BoroughPeter, Boson, Bowman17, Bozna, Bradeos Graphon, Brahmaputra, Brain40, Brainbark, Branddobbe, Brandmeister, Brandmeister (old), Brandt1, Breathe, BrendelSignature, Breno, Brian0918, Brighterorange, Brion VIBBER, Britannicus, BritishWatcher, Brockert, BrokenSegue, Brossow, Bruceknibbs, Brumski, Brutaldeluxe, Brutannica, Bryan Derksen, Bssc81, Buaidh, Bubba hotep, Buchanan-Hermit, Buddha24, BumHoleo, Bunny-chan, Burpelson AFB, Bushy moustache, BuzzWoof, Bwithh, ByrnedHead, CJ Aikman, CORNELIUSSEON, CSWarren, Cactus.man, Caerwine, Cakeandicecream, CalJW, CalendarWatcher, CalicoCatLover, Caltas, Calum MacÚisdean, Cam275, CambridgeBayWeather, Cambyses, Cameltoe237120236419324, Camilton, Can't sleep, clown will eat me, CanDo, Canadian-Bacon, Canaen, Canationalist, Canderra, Caniago, Canterbury Tail, Cantus, Captain Courageous, Captain Future, CaptainUnderpants, CaptainVindaloo, Captainspizzo, Carabinieri, Carifio24, CarlCarlton, Carlossuarez46, Carnby, Carson56437, Caryptes, Casper2k3, Catalyst in Society, Catdude, Causa sui, Cause of death, Cautious, Caveman 07, Cedrus-Libani, Celestra, Celyndel, Cenarium, Censudata, Centrx, Cethegus, Ch'marr, ChKa, Chaddy, Chairboy, Chalst, Chanting Fox, Chaosdruid, CharlesMartel, Charly33, CharonX, Chatanga, Che829, Checkmate987, Chederman69, Cheeseheadxvi, Chicheley, Childzy, Chillum, Chipmunkdavis, Chochopk, Chodorkovskiy, Chokerman88, Choop, Choster, Chowbok, Chris 73, Chris Archer, Chris the speller, Chrisfow, Chrisritch, Christian List, Christinam, Chriswiki, Chrysalis, Chtrede, Chubbles, Chuck Carroll, Chucky21, Church of emacs, Chuunen Baka, Chwyatt, Chzz, Ciacchi, Ciceronl, Ciddler, CieloEstrellado, Cirt, Citizen-of-wiki, City-17, Civil Engineer III, CjDMaX, Cjthellama, Ckatz, Ckorff, Classical 2006, Cliché Online, Cmathio, Cmdrjameson, Coasterlover1994, Coat of Arms, Codycod10, Coffeinfreak, Cogiati, Cogito ergo sumo, Cognate247, Cokes360, Colchicum, ColdWind, Colonies Chris, CommonsDelinker, ConWilBur, Conor's Whipped, Conorbrady.ie, Conorobradaigh, Conti, Contributor777, Conversion script, Cooksey, Coolcole93, Cordless Larry, Coredesat, Cosmic Latte, Crabbit, Craigzomack, Crazycomputers, Crazycoolalex, CrimsonBlue, Crisskriss, Crissov, Crito2161, CrnaGora, Crossk, Crowbar1234, Crusadeonilliteracy, Cruzian, Cryptic, Crystallina, Crzrussian, Cs-wolves, Csörföly D, Cukiger, Culveyhouse, Curps, Curtisvic, Cvieg, Cyde, Cyfal, Cyopardi, Cyrius, D0rkypn0y1029, D34d10gic, D6, DARTH SIDIOUS 2, DCGeist, DGJM, DJ Clayworth, DLiebisch, DMacks, DO'Neil, DTGHYUKLPOQWMNB, DTOx, DVD R W, DaL33T, DaQuirin,

Dakhart, Dalmatian Mommy, Damian Radu, Dan D. Ric, Dan Guan, Dan100, DanMS, Dana boomer, Danaman5, Danc, DancingPenguin, Daniel5127, DanielCD, Danielt, Danny, Dantheman531, Dapete, Daqui, Darealclub, DarkFalls, Darkieboy236, Dasani, Dave031970, Dave420, Daveb, Davesmall, David J. Rogers, David Kernow, David Levy, David Liuzzo, David N. Hermann, David.Mestel, Davidfollsom, Davidpdx, Dbachmann, Dca5347, Dcandeto, Dcflyer, Deadblob93, Deanos, Debate789, Debresser, Dedeler31, Deejaye6, Deflective, DefunKt, Deiz, Delldot, Deltabeignet, Demiurge, Democryt, Demomoke, Dendodge, DennisDaniels, DennisGigio, Denniss, Deppp, Der Eberswalder, Der Statistiker, Der Wolf im Wald, DerHexer, Deszcz, Deutsche, DevastatorIIC, Devatipan, Deviathan, Deville, Dewritech, Dfrg.msc, Dgg32, Dhp1080, DickLukensTheSecond, Dierk, Dinner123, DirectorR, Directorstratton, DirkvdM, Dirtyhose, DisappointedHorse, Discospinster, Dispenser, DivineIntervention, Dizzyizzy, Djmutex, Dkusic1, Dlugopis, DocendoDiscimus, Doco, Doctor Allen, Docu, Dogface, Dojarca, Donnamaria, Donnerstag, DopefishJustin, Dorftrottel, Doric Loon, Doug Johnson, Dowj, Downing Street, Dpr, Dr. Crisp, Dr.alf, DrBob, DrKiernan, Dragoburaggo, Dragoon009, Drewbean, Drewrau, Drgals, Driftwood87, Drmies, Drunkenmonkey, Dub4u, Dude8679, Dukeofomnium, Duncharris, Durin, DutchmanJoy, Dysmorodrepanis, Dzenanz, E Pluribus Anthony, E-boy, E235, EBB, ESkog, Easter Monkey, Eatyourgreens, Eco84, Eddospants, Eden Tate, Edgar181, Editor at Large, Editor18, Edwy, EffK, Ehrenkater, Eirik.lingas, Eixo, Ejgreen77, Ekki01, Eklipse, El C, Electionworld, Elekhh, Eleuther, Elf-friend, Elfguy, Elizabeth A, Elliskev, Ellmist, Elmondo21st, Elockid, Eloquence, Elvis, Emperorbma, EncycloPetey, Endofskull, Endurance, EnemyOfTheState, Energyfreezer, Engleaugen6666, Englishnerd, Enlil Ninlil, Enviroboy, EoGuy, Eon, Er Komandante, Eraliswelsh, Erauch, Eric Yurken, EricS, Ericoides, Erlando, ErnieBern, Erstats, Erudra, Esanchez7587, Escape Artist Swyer, EscapingLife, Eschbaumer, Esimal, Etams, Eu.stefan, Euchiasmus, Eugen Simion 14, Eugene van der Pijll, Everyking, Evice, Evil Monkey, F2020, FAThomssen, FF2010, Fabartus, Factuarius, Fahima07, Fainites, Fair Deal, Fallschirmjäger, Fama21, FamilyGuy1998, Familyrits, Farooqmoazzam, Fastestdogever, FatM1ke, Favonian, Fawhad, Feldfrei, Feldmarschall, Felixboy, Ferrija1, Ferro Carlotta Monzi Brak, Fheidener, Fidelio72, Fiet Nam, Filelakeshoe, Finalius, Finlay, FinnishDriver, Firetrap9254, Firsfron, Fisss, Fkbreitl, Flammingo, Flappychappy, Flatterworld, Flauto Dolce, Florentino floro, Florian Blaschke, Flowanda, Flowerpotman, Fnfd, Fonzy, Fooloomanzoo, ForOurXylophones, FordPrkt, Forstnera, Fossa, Fotaun, Four-ArmedRevenge, Fourtimesfour, Franciswiki, Francs2000, Frank Anchor, Freak in the bunnysuit, Freakincoursework, Freakofnurture, Fred Stober, Freddo207, FreplySpang, Friedrichshainer, FritzK0310, Frymaster, Fschoenm, Fsotrain09, Fudge43, Fudoreaper, Fujishima, Fulcher, Funkelnagelneu, Funnyhat, Furrykef, Future Perfect at Sunrise, Fuzheado, Fuzzy510, Fvincent, Fvw, Fwappler, Fyodor, G-Man, G. Campbell, GBWallenstein, GHe, GLGerman, Gabbe, Gabithefirst, Gabriel-Royce, Gadfium, Gaius Cornelius, Gaius Octavius Princeps, Galati, Galoubet, Gamer007, Garryq, Gausie jr, Gavia immer, Gavin Moodie, Gaz, Gekritzl, GenQuest, Geneb1955, Geni, Gennarous, Geo-Loge, Georg Hurtig, GeorgeTSLC, GeraldH, Gerhard51, German Food, German name, GermanJoe, Germanclub, Germanuser, Gert7, Gertjan R., Gertyness, Getker, Gfoley4, Gggh, Ggoere, Ghepeu, GiZiBoNG, Gidonb, Gilbertfein, Gilgamesh, Gilliam, Gilo1969, Gimmetrow, Giovanni76, Girdi, Giving is good, as long as you're getting, Gjp23, Glen, Glenn, Gmaxwell, Gmosoti, GnuDoyng, Godefroy, GoetheFromm, Gogafax, Gogo Dodo, Golbez, Good Olfactory, GoodSirJava, Goodoldpolonius2, Googl, Googleguy1234, Gotrl, Gr8opinionater, Grace Note, Gracenotes, Grade7, GraemeL, Grafen, Grafikm fr, Graham87, Gran2, Graufbold, Greatal386, Greatgreenwhale, GreekNurse, Green Giant, Green Politics, Green caterpillar, GreenUpGreenOut, GregAsche, Grendelkhan, Greyhood, Griffinofwales, GrimmC, Grinman, Grmpf, Gronau, Grstain, Grunt, Gryffindor, Gsklee, Gtg204y, Guanaco, Gugganij, Guidedbyalan, Guideended1, Guilherme Paula, Guliologez, Gunnar.Kaestle, Guoguo12, Gurch, Gurchzilla, Guy M, Gwernol, Gwyon, Gzornenplatz, H-stt, H0ttamale2, H1523702, HJ Mitchell, HVH, Hadal, Hadseys, HagSibylle, Haham hanuka, Hairy Dude, Hajduk, Hajor, Half price, HalfShadow, Halibutt, Hall Monitor, HamboLSSAH, Hannil, HappyCamper, Harmil, Harryboyles, Hasan Dalbudak, Hassocks5489, Hatt, Haukurth, Hawkestone, Hayden120, Haymaker, Hazzl, Hdante, Heah, Heimdal, Heimstern, Helsinkicommission, Helwerda, Hemanshu, Henry Flower, HerkusMonte, Heron, Herostratus, Herr Kent, Hesus321, Hexil blitz, Hhielscher, Highfields, Highpoems123, Hishii, Historian19, Hmains, Hmwith, Hoc30, Holonboy, Holy Oil And Water, Holynightfever, Homagetocatalonia, Homerjay, Honette, Hoo man, HoodedMan, Hornymanatee, Horst-schlaemma, Househelps, Hqb, Hugo999, HungarianWoman, Husamelodeh, Husond, Hydrargyrum, Hydraton31, I already forgot, IJA, IJzeren Jan, Ian Pitchford, Ibn Battuta, IceArchon, Icecradle, Ich, Ich-AG, Idaltu, Ideletealot2, Ief, Igo4U, Iksisy, Ilehmann, Ilgom, Illythr, Iluvcapra, Ilya1166, Imadcow1, Imaglang, Imlepid, Imperator, ImperatorExercitus, Imperium Europeum, Improv, Imtrogdor, Imwivstuipid, InShaneee, Indon, Infamouskitty, Information-Line, Inge-Lyubov, Ingebjorg, Ingolfson, Instantnood, Int21h, Inwind, Iol, Iowausa, Ipsofactoid, Irdepesca572, Iridescent, Irishguy, Iritebs222, IronGargoyle, Irrypride, IsarSteve, Istabraq, It Is Me Here, Italiano111, Ivirivi00, Ixfd64, Izehar, Izzedine, J Di, J Milburn, J. 'mach' wust, J.K Rowling, J.delanoy, J00tel, JAU123, JFT96, JForget, JHMM13, JLogan, JM.Beaubourg, JMS123, JNW, JTBX, Jacek Kendysz, Jack Silver1000, Jackobyte, Jacques2, Jadger, Jaellee, Jaggsan, Jajafe, Jakobburkhardt, James086, James297, Jamesooders, Jamyskis, Jan Hofmann, Janneman, Jansch, Jared Preston, Jarjarbinks10, Jarkeld, Jasman, Jatkins, Java13690, Java7837, Jaxl, Jaxsonjo, Jayden54, Jayjg, Jazz+, Jbattersby, Jbou, Jcornelius, Jcrook1987, Jcw69, JdH, JdeJ, JeLuF, Jeff G., Jeff3000, Jeffrey O. Gustafson, Jekki, Jeltz, Jensboot, JeremyA, Jeronimo, Jerry67, Jessejamesandloop, Jessica Gordon, Jesushaces, Jesusjonez, Jez, Jfruh, Jhendin, Jiang, Jim bob bryar, Jimiloughlin, Jimp, Jjhcap99, Jjk82, Jklin, Jknabe, Jmcc150, Jmh, JoB614, Joachim Weckermann, JoanneB, JodyB, Joe Canuck, Johan Magnus, Johann Wolfgang, Johannes H.-Larsen, Johanneswilm, John, John D., John K, John R Murray, John Riemann Soong, John Smith M.D., Ph.D, John Smith's, John254, Johnny Undead, Johnpro, Jomika, JonGwynne, JonHarder, JonRoma, Jonah474747, Jonathan Groß, Jondel, Jondumitru, Jonesy, Jonophan, Jooler, Jor, Jose77, Joseph Solis in Australia, Joshua, Jossi, Joséwell, Joy, Jpeob, Jpgordon, Jr724, Jrcrin001, Jrleighton, Jrtsstfjsfkyjsrt6istk7stkyskjytrsk, Jsbhavsar, Jsc83, JuJube, Jujutacular, Julius1990, Jumbuck, Jumpytoo, Junes, Junkietom08, Jusdafax, Jusjih, JustPhil, Justin Eiler, Justin Piga, Justinc, Justinhful, Jwestbrook, Jwordsmith, Jwy, Jyusin, Jza84, KFP, KG200, KJohansson, KMileSupernova, Kacen, Kadillia, Kafziel, KagamiNoMiko, Kahusi, Kaihsu, Kaiser Torikka, Kaiserjd, Kajzderski, Kakofonous, Kalaha, Kamyra, Kantianer, Karam.Anthony.K, Karl, Karl-Henner, KarlFrei, KarlMathiessen, Katherz, KathrynLybarger, Katzensprung, Kbdank71, Kchishol1970, Keegan, Keesiewonder, Keichwa, Keilana, Keith D, Keizuko, Kejaja, Kelly Martin, Kelnor, Kermecke, Kevin, KevinCarmody, KevinOB, KevinTR, Kevs, Kevtrice, Kewp, Kewpid, Khanada, Khoikhoi, Khukri, Killdevil, Killer.k09, KillerChihuahua, Kim-Zhang-Hong, King of Hearts, Kiwifruit89, Kkm010, Klaus Bells, Klausness, Kleuske, Klosterdev, Kman543210, Kman665, Knarf-bz, Knipptang, KnowledgeOfSelf, Knowledgebycoop, Koavf, Komet, Kompakt, Konstantin, Koro lord, Kosebamse, Kotniski, Koyaanis Qatsi, Kozuch, Kr1st1deejay97, KrakatoaKatie, Krankman, Kransky, Krellis, Kribbeh, Kross, Ksenon, Kt66, Kungfuadam, KuroiShiroi, Kuru, Kusma, Kwamikagami, Kwesiidun91, Kwiki, Kyle Benge, Kyle4950, Kyrin, Körnerbrötchen, Kızıl Şaman, L'Etat c'est moi, L1A1 FAL, L95slovenia, LARS, LIU, LKDD, LLLex, LSLM, LaMenta3, Laca, Lacrimosus, LadnaV, Lafuzion, Lakshmix, Lan360, Lancevortex, Landroo, Langston, Larko3344, Larry Sanger, Larry_Sanger, Lars T., Lars Washington, LarsJanZeeuwRules, Laugreve, Laurens-af, Laurentleap, Lawl1232323, Le Fou, Leafyplant, Leandrod, Lear 21, Learngernanwords, Lectonar, Lee J Haywood, Lee S. Svoboda, Leewilson123, Leibniz, Lemsi, Lemsus, Leo Bauer, LeoO3, Lethe, Leuko, Levalley, Levdave7381, Levineps, Levonscott, Lexicon, Lfh, Lhademmor, Lhasaq, Liamdaly620, Liberatus, Lietzow, Liface, LightSpectra, Lightdarkness, Lightmouse, Ligulem, LilHelpa, Lilac Soul, Lilshimon, Lilybaeum, Linuxbeak, Lir, ListedRenegade, LiterallySimon, Lithpiperpilot, LittleDan, LittleOldMe old, LittleRoughRhinestone, LlywelynII, Llywrch, Lobsterantlers, Locke Cole, Lockesdonkey, Logan, Lolhilol, Lolitsjoel16, Lolol123, LonelyMarble, Lonerville, Longbow4u, Longhair, Lord Vader, LordofHavoc, Lordvaluemart, Loren36, Loshaisamazing, Lotje, Louis3ham, Lowellian, Lrobb95, Lstipakov, Lucas323, Lucyin, Ludger1961, Ludjer, Lugnad, LuigiManiac, Luis rib, Luisztdt, Lukatheist, Luna Santin, Lunacy.Fringe, Lunar2, Lunarboy, Lupo, Lurker, Lyfeizshrt, Lysy, MECU, MER-C, MIKElovesgermanchicks, MJCdetroit, MKers06, MKoltnow, MOMBO BAGGINS, MONGO, MPS, MPerel, MTWEmperor, MaGioZal, Macintosh User, Mackensen, Mackeriv, Mactruth, Madman, Maelnuneb, Magicjigpipe, Magister Mathematicae, Magnus Manske, Magumbaboy, Mailer diablo, Malarious, Malbi, Malcolm Farmer, Male1979, Malerin, Malfidus, Malo, Man vyi, Mandavi, Manecke, Maneus4, Mangoe, Manop, Manpyo, MapsMan, Marcika, Marco polo, Marcus Cyron, Marcworca, Marek69, MarekF, Maria Stella, Marielleh, Mark Ryan, Mark.P.Bartlett, MarkThomas, Markus Schmaus, Marky-Son, Marlight, Marnanel, Marriex, MarsRover, Martg76, Martin.schoenberg, Martinp23, Martinwilke1980, Maru4u, Marvan, Marysunshine, Marzoregulus@yahoo.co.ph, Marzter6, Maskan, Massimshowalter, Massjit, Master Jay, MasterLycidas, Materialscientist, MathFacts, Mathew123, MathiasFox, Matilda, Matros, Matstone99, Matt 989, Matt Crypto, Matt512, Mattbaileyisneat, Matthead, Matthew Fennell, Matthewjw 16, Matthäus Wander, Matticus78, Mattingly23, MatttK, Mauriziotani, Mav, Maverick starstrider, Mawa, MaxBech1975, Maximus Rex, Maxschmelling, Maymose, Mayonnaise, Mazzelino, McDogm, McGeddon, Mcewan, Mclay1, Mcmoor, Mcurmamma, Me Mighty Leper, Meelar, MegA, Meh31, Mehmeda, MeinGoshen, Melaen, Mellum, MementoVivere, Menameischarle, Menchi, Mentifisto, Merbabu, Merchbow, Mercury, Merlincooper, Mertan, Mervyn, Mesgul82, Metaljaws, Mfa fariz, Mgiganteus1, Mgoodyear, MiLo28, Miaers, Mic, Mich112358, Michael Hardy, Michael Zimmermann, Michael riber jorgensen, MichaelDiederich, Michaelas10, Michaelmcneil, Michaeltobbylee, Michbich, Michealkelly, Michel rockstar, Mickey gfss2007, Microsomen, Middleeast, Midnightblueowl, Mifter, Mightymights, Miguel Andrade, Miguel.mateo, Mike Rosoft, Mike.heckmann, MikeAtari, Mikeifc123, MillaSukel, Milton Stanley, Mipago, Miroslawa, MisfitToys, Miskin, Missspencer06, Mister C4, MisterSheik, Misza13, Mittermeier, Mixcoatl, Miyokan, Mjpieters, Mmunger, Mnemeson, Mnmazur, Mo-Al, Mo0, Modelun88, Moe Epsilon, Moehre, Mollemy, Molobo, Moncrief, Monkeyknife, Monomachus, Moominjosh, Moonunit02, Mooseberry, Moriori, Morse thor, Morwen, Moshe Constantine Hassan Al-Silverburg, Mouvement, Movementarian, Moxdude, Moxy, Mqduck, Mr Adequate, Mr MillA, Mr Stephen, Mr Woman, Mr. Know-It-All, Mr. bwn, MrCanadian1, MrFish, MrPMonday, MrRandomPerson, MrTranscript, Mramz88, Mrfathead, Mrjeff, Mschel, Mshecket, Mskyfor848, Mtl 16, Mtmoore321, Mucky Duck, Mufka, Munci, Mus Musculus, Muskydusky, Mxcatania, Mxn, My76Strat, Myanw, Mycsrnnm, N-edits, N5iln, NB-NB, NYSTEX1992, Naive cynic, Nakon, NameIsRon, Narya, Naryathegreat, Nascar1996, Natalie Erin, Naverno, NawlinWiki, Nbatra, Nebauer, Nectarflowed, Nellov5, Neozoon, Nero the second, NeroN BG, NetherlandishYankee, Netknowle, Netsnipe, Neurowin, Neutrality, Nevfennas, NewYork483, Newkai, Newnoise, Nicander, Nichalp, Nick San, Nicketje666, Nickinny, Nickispeaki, Nickj, Nicklovesgold, Nicksukh, Nico, Nicofs, Nicolas1995, Nieding, NigelR, Night w, Nihil novi, Nika, Nikkimaria, Niko226, Nikola Smolenski, Nikostar, Nimbusania, Nirvana888, Nishkid64, Nitya Dharma, Nivix, Nixdorf, Njecoolguy90, Nlnnet, Nmpenguin, NoGringo, NoPuzzleStranger, Nobleeagle, Nohomers48, Nolelover, Noonpawit, Noon, Noozgroop, Nopetro, Northernmonkey, Nothlit, Notuncurious, Nouse4aname, Npovshark, Nricardo, Nscheffey, Ntwarrior, NuclearVacuum, NuclearWarfare, Nufy8, Numbo3, Nunquam Dormio, Nv8200p, Nzpcmad, OOODDD, OSX, Obarskyr, Octane, Od Mishehu, OettingerCroat, Ohconfucius, Ohnoitsjamie, Okichubiness, Okinawadude, Olaf Eriksson, Old Guard, Oldnag85, Olegwiki, Olessi, Olgaru, OliFer, OliverH, OlliffeObscurity, Olorin28, Olowek, Oman66, Omicronpersei8, Onco p53, OneGuy, Oneearth, Onlineinteraction, Oo64eva, Oosoom, Opc123, Opelio, Opus88888, Orangerider, OrbitOne, Oreo Priest, Ori.livneh, Ornithologician, Orz, Ospalh, Ot, Ottawa4ever, Otto ter Haar, OwenX, P-Schmidt22, P.juka, PFHLai, PS2pcGAMER, PSWG1920, Paddyez, Paine Ellsworth, Paintball12345, Pakhomovru, Palica, Pan Dan, Panoramastitcher, Pantherskin, Paradocks2, Paranoid, Pascal.Tesson, Pathless, Patka, Patrick, Patstuart, Patyk, Paul Benjamin Austin, Paul Erik, PaulBommel, Pavlo Chemist, Pawel z Niepolomic, Pax:Vobiscum, Pearle, Pecher, Penmachine, Pensionero, Pepsidrinka, Per Abrahamsen, Peter, Peter Horn, Peter Isotalo, Peter2005, PeterCS, Peterale4444, Pethr, Petree13, Petri Krohn, Petrux, Pewwer42, Pgk, Phaedriel, Phantomsteve, Pharaoh of the Wizards, Phatom87, Phenz, Phil Bastian, Phil Boswell, PhilKnight, Philip Baird Shearer, Philip Cross, Philip ea, Phillip J, Philophobia, Philter137, Phoenix of9, Picapica, Picaroon, Piccadilly, Piggyspizza, Pilotguy, Piotrus, Piratelooksat30some2001, Pirawik, Pjacobi, Platane, Plocho22, Plumber, Pofka, Pokrajac, Polaron, Polylepsis, Pommerland, Popadopolis, Porqin, Porsche997SBS, Postcard Cathy, PostdIf, Postmann Michael, Prattlement, Preslethe, Primate, PrimeCupEevee, Pristino, ProhibitOnions, Proteus, Prumpf, Pschemp, Pseudo-Richard, PseudoSudo, Psg116, Pslide, Psy guy, Puchiko, Punctilius, Purple Sheep, Puttputttime, PuzzletChung, Pyroblondie, Qelp, Qero, Quakenul, QuartierLatin1968, QueenStupid, Queson, Quibik, Quizimodo, Quizimodo (usurped), Qweedc, Qwerty Binary, R Lowry, R'n'B, R-41, R9tgokunks, RG2, RHB, RM21, RMHED, RPlunk, RSido, Rada, Radagast, Radu Gherasim, RaffiKojian, RagingR2, Railer 353, Ralphy, Ran, Randomblue, RandySavageFTW, RapaNui75, Rarelibra, Raul14, Raul654, Raven4x4x, Ray Van De Walker, Raymond arritt, Rbz321, Rd232, Reboot, Reconsider the static, Recury, Red Winged Duck, RedWolf, Reddi, Redking7, Rednaxela, Reedy, Reimanova, Reindra, Reinis, Reisio, Rejedef, Renata, ResearchRave, Resilienzi, Retired username, Retireduser1111, Rettetast, Revolving Bugbear, Rex Germanus, RexNL, Reywas92, Rflesher, Rhys fry, Riana, Ribbet32, Ricbep, Rich Farmbrough, Richard D. LeCour, Richardcavell, Richarddd, Richardfabi, Rick Block, RickK, Ricky@36, Rickyrab, Rilekou, Rimshot, Rje, Rjensen, Rjwilmsi, Rl, Rmky87, Rmt2m, Rnelson, RoMex, RoadTrain, Rob 06-07, Rob.derosa, Robdurbar, Robendo, Robert1947, Robert25, RobertG, Robertsteadman, Robinstocks, Robomaeyhem, Robzz, Rocastelo, RockMFR, Rocket000, Rockfang, Rockstar4000, Rod57, Rogerd, Rokbas, Romankawe, Romanm, Ronhjones, Rosenzweig, Routerrandy4, Roxi2, Roy da Vinci, RoyBoy, Royboycrashfan, Rracecarr, Rreagan007, RuffRyderMohan, Ruhrjung, Rurururu, RussBlau, Ruud64, RxS, RyanGerbil10, Ryulong, SFC9394, SKC, SNIyer12, Sacxpert, Saforrest, Sagaciousuk, Said 13, Saimdusan,

Image Sources, Licenses and Contributors

File:Wappen_Waltenhausen.png *Source*: http://en.wikipedia.org/w/index.php?title=File:Wappen_Waltenhausen.png *License*: unknown *Contributors*: User:Lou.gruber

file:Germany location map.svg *Source*: http://en.wikipedia.org/w/index.php?title=File:Germany_location_map.svg *License*: unknown *Contributors*: User:NordNordWest

File:Red_pog.svg *Source*: http://en.wikipedia.org/w/index.php?title=File:Red_pog.svg *License*: unknown *Contributors*: Anomie

Image:Administrative Gliederung Deutschlands.png *Source*: http://en.wikipedia.org/w/index.php?title=File:Administrative_Gliederung_Deutschlands.png *License*: unknown *Contributors*: User:David Liuzzo

File:Administrative Districts of Germany 1981-2008.svg *Source*: http://en.wikipedia.org/w/index.php?title=File:Administrative_Districts_of_Germany_1981-2008.svg *License*: unknown *Contributors*: Bennet Schulte

File:Wappen_Bad_Wiessee.jpg *Source*: http://en.wikipedia.org/w/index.php?title=File:Wappen_Bad_Wiessee.jpg *License*: unknown *Contributors*: Dominix, Magul, Moros, Rauenstein, Redline is courtage

Image:Bad Wiessee Maria Himmelfahrt.jpg *Source*: http://en.wikipedia.org/w/index.php?title=File:Bad_Wiessee_Maria_Himmelfahrt.jpg *License*: unknown *Contributors*: User:Rufus46

File:Bavaria mb.png *Source*: http://en.wikipedia.org/w/index.php?title=File:Bavaria_mb.png *License*: unknown *Contributors*: Jed, Moros, TUBS

File:Flag of Germany.svg *Source*: http://en.wikipedia.org/w/index.php?title=File:Flag_of_Germany.svg *License*: unknown *Contributors*: Anomie

image:Wappen Landkreis Miesbach.png *Source*: http://en.wikipedia.org/w/index.php?title=File:Wappen_Landkreis_Miesbach.png *License*: unknown *Contributors*: User:Rosenzweig

File:Flag of Bavaria (lozengy).svg *Source*: http://en.wikipedia.org/w/index.php?title=File:Flag_of_Bavaria_(lozengy).svg *License*: unknown *Contributors*: diese Datei: Jwnabd

File:Coat of arms of Bavaria.svg *Source*: http://en.wikipedia.org/w/index.php?title=File:Coat_of_arms_of_Bavaria.svg *License*: unknown *Contributors*: Freistaat Bayern

File:Deutschland Lage von Bayern.svg *Source*: http://en.wikipedia.org/w/index.php?title=File:Deutschland_Lage_von_Bayern.svg *License*: unknown *Contributors*: User:David Liuzzo

File:Speaker Icon.svg *Source*: http://en.wikipedia.org/w/index.php?title=File:Speaker_Icon.svg *License*: unknown *Contributors*: Blast, G.Hagedorn, Mobius, 2 anonymous edits

File:Heunischenburg.jpg *Source*: http://en.wikipedia.org/w/index.php?title=File:Heunischenburg.jpg *License*: unknown *Contributors*: Stefan Wicklein, Kronach

File:Bayern im 10.Jh.png *Source*: http://en.wikipedia.org/w/index.php?title=File:Bayern_im_10.Jh.png *License*: unknown *Contributors*: Electionworld, Ephraim33, Koryakov Yuri, Rfortner, 1 anonymous edits

File:Bayern nach der Teilung 1392.png *Source*: http://en.wikipedia.org/w/index.php?title=File:Bayern_nach_der_Teilung_1392.png *License*: unknown *Contributors*: User:Lencer

File:Bayern von 1800 bis heute.png *Source*: http://en.wikipedia.org/w/index.php?title=File:Bayern_von_1800_bis_heute.png *License*: unknown *Contributors*: Original uploader was Tk at de.wikipedia

File:Block of Bavarian stamps (1920s) overprinted with "Deutsches Reich".jpg *Source*: http://en.wikipedia.org/w/index.php?title=File:Block_of_Bavarian_stamps_(1920s)_overprinted_with_"Deutsches_Reich".jpg *License*: unknown *Contributors*: User:Atzmonit

File:Kriegerdenkmal Dietelskirchen.jpg *Source*: http://en.wikipedia.org/w/index.php?title=File:Kriegerdenkmal_Dietelskirchen.jpg *License*: unknown *Contributors*: Seppal

File:Bavarian Herald.jpg *Source*: http://en.wikipedia.org/w/index.php?title=File:Bavarian_Herald.jpg *License*: unknown *Contributors*: Evadb, FordPrefect42, Mattes, Mogelzahn, 2 anonymous edits

File:Bavarian Alps 2002.jpg *Source*: http://en.wikipedia.org/w/index.php?title=File:Bavarian_Alps_2002.jpg *License*: unknown *Contributors*: Melancholia i

File:Bavarian Admin Districts.jpg *Source*: http://en.wikipedia.org/w/index.php?title=File:Bavarian_Admin_Districts.jpg *License*: unknown *Contributors*: Pere prlpz, TUBS

File:Wappen_Oberbayern.svg *Source*: http://en.wikipedia.org/w/index.php?title=File:Wappen_Oberbayern.svg *License*: unknown *Contributors*: Freistaat Bayern

File:Wappen Bezirk Niederbayern.svg *Source*: http://en.wikipedia.org/w/index.php?title=File:Wappen_Bezirk_Niederbayern.svg *License*: unknown *Contributors*: User:Flow2

File:Wappen_Operpfalz.svg *Source*: http://en.wikipedia.org/w/index.php?title=File:Wappen_Operpfalz.svg *License*: unknown *Contributors*: Freistaat Bayern

File:Wappen Bezirk Oberfranken2.svg *Source*: http://en.wikipedia.org/w/index.php?title=File:Wappen_Bezirk_Oberfranken2.svg *License*: unknown *Contributors*: peppeg

File:Mittelfranken_Wappen.svg *Source*: http://en.wikipedia.org/w/index.php?title=File:Mittelfranken_Wappen.svg *License*: unknown *Contributors*: User:Beck.worma

File:Unterfranken Wappen.svg *Source*: http://en.wikipedia.org/w/index.php?title=File:Unterfranken_Wappen.svg *License*: unknown *Contributors*: Thomas Büttner

File:Wappen_Schwaben_Bayern.svg *Source*: http://en.wikipedia.org/w/index.php?title=File:Wappen_Schwaben_Bayern.svg *License*: unknown *Contributors*: User:Flow2

File:Karte-Bayern-Landkreise.png *Source*: http://en.wikipedia.org/w/index.php?title=File:Karte-Bayern-Landkreise.png *License*: unknown *Contributors*: User Cordyph on en.wikipedia

File:Rathaus and Marienplatz from Peterskirche - August 2006.jpg *Source*: http://en.wikipedia.org/w/index.php?title=File:Rathaus_and_Marienplatz_from_Peterskirche_-_August_2006.jpg *License*: unknown *Contributors*: User:Diliff

File:Nürnberg panorama.jpg *Source*: http://en.wikipedia.org/w/index.php?title=File:Nürnberg_panorama.jpg *License*: unknown *Contributors*: 34er, AndreasPraefcke, BlackIceNRW, Mb1302, Technokrat

File:Rathaus Augsburg.jpg *Source*: http://en.wikipedia.org/w/index.php?title=File:Rathaus_Augsburg.jpg *License*: unknown *Contributors*: ElRaki, Fransvannes, Gryffindor, Hullbr3ach, Ronaldino, Till.niermann, 2 anonymous edits

File:Stadtansicht Regensburg.JPG *Source*: http://en.wikipedia.org/w/index.php?title=File:Stadtansicht_Regensburg.JPG *License*: unknown *Contributors*: HH58

File:Würzburg view.jpg *Source*: http://en.wikipedia.org/w/index.php?title=File:Würzburg_view.jpg *License*: unknown *Contributors*: Burny, FlickrLickr, Ronaldino, Till.niermann, Torben

File:Marktplatz Bayreuth.JPG *Source*: http://en.wikipedia.org/w/index.php?title=File:Marktplatz_Bayreuth.JPG *License*: unknown *Contributors*: User:Michael Sander

File:Bamberg Altes Rathaus BW 1.JPG *Source*: http://en.wikipedia.org/w/index.php?title=File:Bamberg_Altes_Rathaus_BW_1.JPG *License*: unknown *Contributors*: User:Berthold Werner

File:Germany bavaria alps-church.jpg *Source*: http://en.wikipedia.org/w/index.php?title=File:Germany_bavaria_alps-church.jpg *License*: unknown *Contributors*: User:RRKennison

File:Karwendel-Ahornboden.jpg *Source*: http://en.wikipedia.org/w/index.php?title=File:Karwendel-Ahornboden.jpg *License*: unknown *Contributors*: Herzi Pinki, Shyam, Svíčková, Zinnmann

File:Kloster_Fürstenfeld_Heilige_Hyacinthus.JPG *Source*: http://en.wikipedia.org/w/index.php?title=File:Kloster_Fürstenfeld_Heilige_Hyacinthus.JPG *License*: unknown *Contributors*: User:Richard Huber

File:Upper German Languages.PNG *Source*: http://en.wikipedia.org/w/index.php?title=File:Upper_German_Languages.PNG *License*: unknown *Contributors*: User:Imperator24

File:Aschaffenburg Schloss Johannisburg.jpg *Source*: http://en.wikipedia.org/w/index.php?title=File:Aschaffenburg_Schloss_Johannisburg.jpg *License*: unknown *Contributors*: User:Steschke

File:Wuerzburger_Residenz_vom_Hofgarten.jpg *Source*: http://en.wikipedia.org/w/index.php?title=File:Wuerzburger_Residenz_vom_Hofgarten.jpg *License*: unknown *Contributors*: Wolfram Esser

File:Marienberg wuerzburg.jpg *Source*: http://en.wikipedia.org/w/index.php?title=File:Marienberg_wuerzburg.jpg *License*: unknown *Contributors*: Christian Horvat

File:Plassenburg oben.jpg *Source*: http://en.wikipedia.org/w/index.php?title=File:Plassenburg_oben.jpg *License*: unknown *Contributors*: Diabas, Moros, Schubbay, Sir Gawain

File:BambergDom.jpg *Source*: http://en.wikipedia.org/w/index.php?title=File:BambergDom.jpg *License*: unknown *Contributors*: AnRo0002, Kjetil r, Kurpfalzbilder.de, Maksim, Para, Schubbay, 1 anonymous edits

File:Vierzehnheiligen I.JPG *Source*: http://en.wikipedia.org/w/index.php?title=File:Vierzehnheiligen_I.JPG *License*: unknown *Contributors*: Asio otus, Ronaldino, Schubbay, Yarl

File:Coburg-Veste1.jpg *Source*: http://en.wikipedia.org/w/index.php?title=File:Coburg-Veste1.jpg *License*: unknown *Contributors*: User:Störfix

File:Bayreuth_Festspielhaus_2006-07-16.jpg *Source*: http://en.wikipedia.org/w/index.php?title=File:Bayreuth_Festspielhaus_2006-07-16.jpg *License*: unknown *Contributors*: Rico Neitzel

File:Nuremberg sebald castle f lorenz f s.jpg *Source*: http://en.wikipedia.org/w/index.php?title=File:Nuremberg_sebald_castle_f_lorenz_f_s.jpg *License*: unknown *Contributors*: User:Keichwa

File:Kastell Biriciana (Weißenburg in Bayern).jpg *Source*: http://en.wikipedia.org/w/index.php?title=File:Kastell_Biriciana_(Weißenburg_in_Bayern).jpg *License*: unknown *Contributors*: Bapti, Brego, Breugelius, Cherubino, Cristiano64, Umherirrender, Xenophon

File:Kreuztor_Ingoldstadt.jpg *Source*: http://en.wikipedia.org/w/index.php?title=File:Kreuztor_Ingoldstadt.jpg *License*: unknown *Contributors*: AnRo0002, Androl, Cresi, Franconia, Fransvannes, Haui, High Contrast

File:Schloss Neuburg.jpg *Source*: http://en.wikipedia.org/w/index.php?title=File:Schloss_Neuburg.jpg *License*: unknown *Contributors*: Original uploader was Mendli at de.wikipedia

File:Einstein 1921 by F Schmutzer.jpg *Source*: http://en.wikipedia.org/w/index.php?title=File:Einstein_1921_by_F_Schmutzer.jpg *License*: unknown *Contributors*: Calliopejen1, DIREKTOR, Frank C. Müller, Hemulen, Lobo, Quibik, Raeky, Андрей Романенко

File:Pyramide Allemagne.PNG *Source*: http://en.wikipedia.org/w/index.php?title=File:Pyramide_Allemagne.PNG *License*: unknown *Contributors*: User:fargomeD

File:Kdom.jpg *Source*: http://en.wikipedia.org/w/index.php?title=File:Kdom.jpg *License*: unknown *Contributors*: ger1axg

File:Heidelberg Universitätsbibliothek 2003.jpg *Source*: http://en.wikipedia.org/w/index.php?title=File:Heidelberg_Universitätsbibliothek_2003.jpg *License*: unknown *Contributors*: Jan Beckendorf

File:Beethoven.jpg *Source*: http://en.wikipedia.org/w/index.php?title=File:Beethoven.jpg *License*: unknown *Contributors*: ADGE, Aaron1a12, Butko, Codeispoetry, CommonsDelinker, Herbythyme, Hotshot977, Howcheng, Ies, Interpretix, Jappalang, Johney, Jusjih, Leonce49, Mattes, Neukoln, Pixel ;-), Shakko, Thuresson, אמסי'סה123, 3 anonymous edits

File:Grimm.jpg *Source*: http://en.wikipedia.org/w/index.php?title=File:Grimm.jpg *License*: unknown *Contributors*: AndreasPraefcke, Anne-Sophie Ofrim, Bukk, Gabor, Hannah, Kresspahl, Léna, Mattes, Nillerdk, Origamiemensch, Phrood, Reinhard Dietrich, Shakko, Yuval Y, 1 anonymous edits

File:Black Forest gateau.jpg *Source*: http://en.wikipedia.org/w/index.php?title=File:Black_Forest_gateau.jpg *License*: unknown *Contributors*: Mikelo

File:Westfalenstadion von oben.jpg *Source*: http://en.wikipedia.org/w/index.php?title=File:Westfalenstadion_von_oben.jpg *License*: unknown *Contributors*: Druffeler, EugeneZelenko, Fuzzy510, Hornymanatee, Mbdortmund, Odin statt Jesus, RHunscher, STBR

File:Openstreetmap logo.svg *Source*: http://en.wikipedia.org/w/index.php?title=File:Openstreetmap_logo.svg *License*: unknown *Contributors*: OpenStreetMap

GNU Free Documentation License Version 1.2, November 2002 Copyright (C) 2000,2001,2002 Free Software Foundation, Inc. 59 Temple Place, Suite 330, Boston, MA 02111-1307 USA Everyone is permitted to copy and distribute verbatim copies of this license document, but changing it is not allowed.

0. PREAMBLE

The purpose of this License is to make a manual, textbook, or other functional and useful document "free" in the sense of freedom: to assure everyone the effective freedom to copy and redistribute it, with or without modifying it, either commercially or noncommercially. Secondarily, this License preserves for the author and publisher a way to get credit for their work, while not being considered responsible for modifications made by others. This License is a kind of "copyleft", which means that derivative works of the document must themselves be free in the same sense. It complements the GNU General Public License, which is a copyleft license designed for free software. We have designed this License in order to use it for manuals for free software, because free software needs free documentation: a free program should come with manuals providing the same freedoms that the software does. But this License is not limited to software manuals; it can be used for any textual work, regardless of subject matter or whether it is published as a printed book. We recommend this License principally for works whose purpose is instruction or reference.

1. APPLICABILITY AND DEFINITIONS

This License applies to any manual or other work, in any medium, that contains a notice placed by the copyright holder saying it can be distributed under the terms of this License. Such a notice grants a world-wide, royalty-free license, unlimited in duration, to use that work under the conditions stated herein. The "Document", below, refers to any such manual or work. Any member of the public is a licensee, and is addressed as "you". You accept the license if you copy, modify or distribute the work in a way requiring permission under copyright law. A "Modified Version" of the Document means any work containing the Document or a portion of it, either copied verbatim, or with modifications and/or translated into another language. A "Secondary Section" is a named appendix or a front-matter section of the Document that deals exclusively with the relationship of the publishers or authors of the Document to the Document's overall subject (or to related matters) and contains nothing that could fall directly within that overall subject. (Thus, if the Document is in part a textbook of mathematics, a Secondary Section may not explain any mathematics.) The relationship could be a matter of historical connection with the subject or with related matters, or of legal, commercial, philosophical, ethical or political position regarding them. The "Invariant Sections" are certain Secondary Sections whose titles are designated, as being those of Invariant Sections, in the notice that says that the Document is released under this License. If a section does not fit the above definition of Secondary then it is not allowed to be designated as Invariant. The Document may contain zero Invariant Sections. If the Document does not identify any Invariant Sections then there are none. The "Cover Texts" are certain short passages of text that are listed, as Front-Cover Texts or Back-Cover Texts, in the notice that says that the Document is released under this License. A Front-Cover Text may be at most 5 words, and a Back-Cover Text may be at most 25 words. A "Transparent" copy of the Document means a machine-readable copy, represented in a format whose specification is available to the general public, that is suitable for revising the document straightforwardly with generic text editors or (for images composed of pixels) generic paint programs or (for drawings) some widely available drawing editor, and that is suitable for input to text formatters or for automatic translation to a variety of formats suitable for input to text formatters. A copy made in an otherwise Transparent file format whose markup, or absence of markup, has been arranged to thwart or discourage subsequent modification by readers is not Transparent. An image format is not Transparent if used for any substantial amount of text. A copy that is not "Transparent" is called "Opaque". Examples of suitable formats for Transparent copies include plain ASCII without markup, Texinfo input format, LaTeX input format, SGML or XML using a publicly available DTD, and standard-conforming simple HTML, PostScript or PDF designed for human modification. Examples of transparent image formats include PNG, XCF and JPG. Opaque formats include proprietary formats that can be read and edited only by proprietary word processors, SGML or XML for which the DTD and/or processing tools are not generally available, and the machine-generated HTML, PostScript or PDF produced by some word processors for output purposes only. The "Title Page" means, for a printed book, the title page itself, plus such following pages as are needed to hold, legibly, the material this License requires to appear in the title page. For works in formats which do not have any title page as such, "Title Page" means the text near the most prominent appearance of the work's title, preceding the beginning of the body of the text. A section "Entitled XYZ" means a named subunit of the Document whose title either is precisely XYZ or contains XYZ in parentheses following text that translates XYZ in another language. (Here XYZ stands for a specific section name mentioned below, such as "Acknowledgements", "Dedications", "Endorsements", or "History".) To "Preserve the Title" of such a section when you modify the Document means that it remains a section "Entitled XYZ" according to this definition. The Document may include Warranty Disclaimers next to the notice which states that this License applies to the Document. These Warranty Disclaimers are considered to be included by reference in this License, but only as regards disclaiming warranties: any other implication that these Warranty Disclaimers may have is void and has no effect on the meaning of this License.

2. VERBATIM COPYING

You may copy and distribute the Document in any medium, either commercially or noncommercially, provided that this license, the copyright notices, and the license notice saying this License applies to the Document are reproduced in all copies, and that you add no other conditions whatsoever to those of this License. You may not use technical measures to obstruct or control the reading or further copying of the copies you make or distribute. However, you may accept compensation in exchange for copies. If you distribute a large enough number of copies you must also follow the conditions in section 3. You may also lend copies, under the same conditions stated above, and you may publicly display copies.

3. COPYING IN QUANTITY

If you publish printed copies (or copies in media that commonly have printed covers) of the Document, numbering more than 100, and the Document's license notice requires Cover Texts, you must enclose the copies in covers that carry, clearly and legibly, all these Cover Texts: Front-Cover Texts on the front cover, and Back-Cover Texts on the back cover. Both covers must also clearly and legibly identify you as the publisher of these copies. The front cover must present the full title with all words of the title equally prominent and visible. You may add other material on the covers in addition. Copying with changes limited to the covers, as long as they preserve the title of the Document and satisfy these conditions, can be treated as verbatim copying in other respects. If the required texts for either cover are too voluminous to fit legibly, you should put the first ones listed (as many as fit reasonably) on the actual cover, and continue the rest onto adjacent pages. If you publish or distribute Opaque copies of the Document numbering more than 100, you must either include a machine-readable Transparent copy along with each Opaque copy, or state in or with each Opaque copy a computer-network location from which the general network-using public has access to download using public-standard network protocols a complete Transparent copy of the Document, free of added material. If you use the latter option, you must take reasonably prudent steps, when you begin distribution of Opaque copies in quantity, to ensure that this Transparent copy will remain thus accessible at the stated location until at least one year after the last time you distribute an Opaque copy (directly or through your agents or retailers) of that edition to the public. It is requested, but not required, that you contact the authors of the Document well before redistributing any large number of copies, to give them a chance to provide you with an updated version of the Document.

4. MODIFICATIONS

You may copy and distribute a Modified Version of the Document under the conditions of sections 2 and 3 above, provided that you release the Modified Version under precisely this License, with the Modified Version filling the role of the Document, thus licensing distribution and modification of the Modified Version to whoever possesses a copy of it. In addition, you must do these things in the Modified Version: A. Use in the Title Page (and on the covers, if any) a title distinct from that of the Document, and from those of previous versions (which should, if there were any, be listed in the History section of the Document). You may use the same title as a previous version if the original publisher of that version gives permission. B. List on the Title Page, as authors, one or more persons or entities responsible for authorship of the modifications in the Modified Version, together with at least five of the principal authors of the Document (all of its principal authors, if it has fewer than five), unless they release you from this requirement. C. State on the Title page the name of the publisher of the Modified Version, as the publisher. D. Preserve all the copyright notices of the Document. E. Add an appropriate copyright notice for your modifications adjacent to the other copyright notices. F. Include, immediately after the copyright notices, a license notice giving the public permission to use the Modified Version under the terms of this License, in the form shown in the Addendum below. G. Preserve in that license notice the full lists of Invariant Sections and required Cover Texts given in the Document's license notice. H. Include an unaltered copy of this License. I. Preserve the section Entitled "History", Preserve its Title, and add to it an item stating at least the title, year, new authors, and publisher of the Modified Version as given on the Title Page. If there is no section Entitled "History" in the Document, create one stating the title, year, authors, and publisher of the Document as given on its Title Page, then add an item describing the Modified Version as stated in the previous sentence. J. Preserve the network location, if any, given in the Document for public access to a Transparent copy of the Document, and likewise the network locations given in the Document for previous versions it was based on. These may be placed in the "History" section. You may omit a network location for a work that was published at least four years before the Document itself, or if the original publisher of the version it refers to gives permission. K. For any section Entitled "Acknowledgements" or "Dedications", Preserve the Title of the section, and preserve in the section all the substance and tone of each of the contributor acknowledgements and/or dedications given therein. L. Preserve all the Invariant Sections of the Document, unaltered in their text and in their titles. Section numbers or the equivalent are not considered part of the section titles. M. Delete any section Entitled "Endorsements". Such a section may not be included in the Modified Version. N. Do not retitle any existing section to be Entitled "Endorsements" or to conflict in title with any Invariant Section. O. Preserve any Warranty Disclaimers. If the Modified Version includes new front-matter sections or appendices that qualify as Secondary Sections and contain no material copied from the Document, you may at your option designate some or all of these sections as invariant. To do this, add their titles to the list of Invariant Sections in the Modified Version's license notice. These titles must be distinct from any other section titles. You may add a section Entitled "Endorsements", provided it contains nothing but endorsements of your Modified Version by various parties--for example, statements of peer review or that the text has been approved by an organization as the authoritative definition of a standard. You may add a passage of up to five words as a Front-Cover Text, and a passage of up to 25 words as a Back-Cover Text, to the end of the list of Cover Texts in the Modified Version. Only one passage of Front-Cover Text and one of Back-Cover Text may be added by (or through arrangements made by) any one entity. If the Document already includes a cover text for the same cover, previously added by you or by arrangement made by the same entity you are acting on behalf of, you may not add another; but you may replace the old one, on explicit permission from the previous publisher that added the old one. The author(s) and publisher(s) of the Document do not by this License give permission to use their names for publicity for or to assert or imply endorsement of any Modified Version.

5. COMBINING DOCUMENTS

You may combine the Document with other documents released under this License, under the terms defined in section 4 above for modified versions, provided that you include in the combination all of the Invariant Sections of all of the original documents, unmodified, and list them all as Invariant Sections of your combined work in its license notice, and that you preserve all their Warranty Disclaimers. The combined work need only contain one copy of this License, and multiple identical Invariant Sections may be replaced with a single copy. If there are multiple Invariant Sections with the same name but different contents, make the title of each such section unique by adding at the end of it, in parentheses, the name of the original author or publisher of that section if known, or else a unique number. Make the same adjustment to the section titles in the list of Invariant Sections in the license notice of the combined work. In the combination, you must combine any sections Entitled "History" in the various original documents, forming one section Entitled "History"; likewise combine any sections Entitled "Acknowledgements", and any sections Entitled "Dedications". You must delete all sections Entitled "Endorsements".

6. COLLECTIONS OF DOCUMENTS

You may make a collection consisting of the Document and other documents released under this License, and replace the individual copies of this License in the various documents with a single copy that is included in the collection, provided that you follow the rules of this License for verbatim copying of each of the documents in all other respects. You may extract a single document from such a collection, and distribute it individually under this License, provided you insert a copy of this License into the extracted document, and follow this License in all other respects regarding verbatim copying of that document.

7. AGGREGATION WITH INDEPENDENT WORKS

A compilation of the Document or its derivatives with other separate and independent documents or works, in or on a volume of a storage or distribution medium, is called an "aggregate" if the copyright resulting from the compilation is not used to limit the legal rights of the compilation's users beyond what the individual works permit. When the Document is included in an aggregate, this License does not apply to the other works in the aggregate which are not themselves derivative works of the Document. If the Cover Text requirement of section 3 is applicable to these copies of the Document, then if the Document is less than one half of the entire aggregate, the Document's Cover Texts may be placed on covers that bracket the Document within the aggregate, or the electronic equivalent of covers if the Document is in electronic form. Otherwise they must appear on printed covers that bracket the whole aggregate.

8. TRANSLATION

Translation is considered a kind of modification, so you may distribute translations of the Document under the terms of section 4. Replacing Invariant Sections with translations requires special permission from their copyright holders, but you may include translations of some or all Invariant Sections in addition to the original versions of these Invariant Sections. You may include a translation of this License, and all the license notices in the Document, and any Warranty Disclaimers, provided that you also include the original English version of this License and the original versions of those notices and disclaimers. In case of a disagreement between the translation and the original version of this License or a notice or disclaimer, the original version will prevail. If a section in the Document is Entitled "Acknowledgements", "Dedications", or "History", the requirement (section 4) to Preserve its Title (section 1) will typically require changing the actual title.

9. TERMINATION

You may not copy, modify, sublicense, or distribute the Document except as expressly provided for under this License. Any other attempt to copy, modify, sublicense or distribute the Document is void, and will automatically terminate your rights under this License. However, parties who have received copies, or rights, from you under this License will not have their licenses terminated so long as such parties remain in full compliance.

10. FUTURE REVISIONS OF THIS LICENSE

The Free Software Foundation may publish new, revised versions of the GNU Free Documentation License from time to time. Such new versions will be similar in spirit to the present version, but may differ in detail to address new problems or concerns. See http://www.gnu.org/copyleft/. Each version of the License is given a distinguishing version number. If the Document specifies that a particular numbered version of this License "or any later version" applies to it, you have the option of following the terms and conditions either of that specified version or of any later version that has been published (not as a draft) by the Free Software Foundation. If the Document does not specify a version number of this License, you may choose any version ever published (not as a draft) by the Free Software Foundation. ADDENDUM: How to use this License for your documents To use this License in a document you have written, include a copy of the License in the document and put the following copyright and license notices just after the title page: Copyright (c) YEAR YOUR NAME. Permission is granted to copy, distribute and/or modify this document under the terms of the GNU Free Documentation License, Version 1.2 or any later version published by the Free Software Foundation; with no Invariant Sections, no Front-Cover Texts, and no Back-Cover Texts. A copy of the license is included in the section entitled "GNU Free Documentation License". If you have Invariant Sections, Front-Cover Texts and Back-Cover Texts, replace the "with...Texts." line with this: with the Invariant Sections being LIST THEIR TITLES, with the Front-Cover Texts being LIST, and with the Back-Cover Texts being LIST. If you have Invariant Sections without Cover Texts, or some other combination of the three, merge those two alternatives to suit the situation. If your document contains nontrivial examples of program code, we recommend releasing these examples in parallel under your choice of free software license, such as the GNU General Public License, to permit their use in free software.

MIX
Papier aus verantwortungsvollen Quellen
Paper from responsible sources
FSC® C105338

Printed by Books on Demand GmbH, Norderstedt / Germany